THE VOICE OF DR. WERNHER VON BRAUN

AN ANTHOLOGY

COMPILED AND EDITED BY
IRENE E. POWELL-WILLHITE

All rights reserved under article two of the Berne Copyright Convention (1971).
We acknowledge the financial support of the Government of Canada through the
Book Publishing Industry Development Program for our publishing activities.
Published by Collector's Guide Publishing Inc., Box 62034,
Burlington, Ontario, Canada, L7R 4K2
Printed and bound in Canada
THE VOICE OF DR. WERNHER VON BRAUN
Irene E. Powell-Willhite
ISBN: 978-1-894959-64-3
ISSN 1496-6921
©2007 Apogee Books

THE VOICE OF DR. WERNHER VON BRAUN

AN ANTHOLOGY

COMPILED AND EDITED BY
IRENE E. POWELL-WILLHITE

An Apogee Books Publication

– CONTENTS –

Chapter 1: The Voice of Wernher von Braun: 1912-1977 7
Chapter 2: Future Development of the Rocket 12
Chapter 3: 1945-1951 17
Chapter 4: We Need a Coordinated Space Program! 23
Chapter 5: Cold War Paranoia & Sputnik I 33
Chapter 6: Across the Space Frontier 34
Chapter 7: The Associated General Contractors of America, January 1959 . . 35
Chapter 8: Youth Faces Tomorrow 40
Chapter 9: Women's Role in a Changing World 44
Chapter 10: Patriotism in the Space Age 46
Chapter 11: Pennsylvania Military College 49
Chapter 12: Greater Boston Chamber of Commerce 53
Chapter 13: Dedication of Goddard Roswell Museum 56
Chapter 14: University of Florida 59
Chapter 15: Space, Its Problems and You 62
Chapter 16: Minnesota Education Association 66
Chapter 17: Tribute to General John Medaris and the ABMA Team 70
Chapter 18: Doing Science 75
Chapter 19: Immortality 78
Chapter 20: National Security Industrial Association 79
Chapter 21: Catch an Exploding Star 84
Chapter 22: Huntsville Ministerial Association 88
Chapter 23: The Making of a Missile 91
Chapter 24: The Meaning of Space Exploration 98
Chapter 25: Boys Scouts of America 102
Chapter 26: Judicial Conference of the U.S. Court of Appeals 103
Chapter 27: German-American Day Festival 109
Chapter 28: SHAPE (Supreme Headquarters Allied Powers Europe/NATO) . . 110
Chapter 29: Fourth National Conference on the Peaceful Uses of Space 132
Chapter 30: French War College: Welcome to Students at MSFC 138
Chapter 31: University of Alabama Huntsville Campaign $750,000 140
Chapter 32: Freedom to Explore 142
Chapter 33: Central Intelligence Agency 145
Chapter 34: The Challenge of the Century 148
Chapter 35: State of Alabama League of Municipalities 152
Chapter 36: Overseas Press Club of America 157
Chapter 37: Lunar Landing Celebration Dinner at Carriage Inn 159
Chapter 38: Civitan International Convention 161
Chapter 39: Religious Implications of Space Exploration: A Personal View . . 166
Chapter 40: National Association of Secondary School Principals 170
Chapter 41: Fall-Out Effects of Megascience 174
Chapter 42: Face-To-Face: 1971 179
Chapter 43: New Answers—New Questions 183
Chapter 44: Benefits From Space Technology 189
Chapter 45: Appalachian Educational Satellite Project 195
Chapter 46: Uses of Space and Space Technology 198
Chapter 47: Responsible Scientific Investigation and Application 203
Appendix: 50th Birthday 228
Notes 230
Bibliography 232

— CHAPTER 1 —

THE VOICE OF WERNHER VON BRAUN: 1912-1977

We are living in a democracy where the will and the mood of the people count. If you want to accomplish something as big as travel into space, you must win the people for your idea. Being diplomatic is necessary, but it is not enough. You have to be filled with a burning desire to bring your idea to life. You must have absolute faith into the righteousness of your cause, and into your final success. In short, you must be a kind of crusader.[1]

PLATO REFERRED TO RHETORIC AS "THE ART OF WINNING THE SOUL BY DISCOURSE." John Locke submitted that rhetoric is "that powerful instrument of error and deceit." Regardless of the definition, speeches, or rhetoric, offer a way for people to effectively present views they deeply care about. Even in our frenzied era of electronic communication, public speaking remains an indispensable vehicle for the expression of ideas. The ability to speak confidently and convincingly in public is an asset to anyone who wants to take an active role in passing his views along.[2] Wernher von Braun was a highly skilled, charismatic speaker, who could communicate with Congressional Committees, at college commencement activities, with ministers, or Boy Scout troops ... anyone who would listen to his message.

Von Braun and his team were sent to El Paso, Texas, at the end of World War II. After a few years in the desert, von Braun decided to go public in pressing for space exploration. An associate, Dr. Adolf Thiel, later told him, "You started to get your thoughts in magazines and through speeches. Some of us longhairs and purists didn't think very highly of that. And one day I opened my big mouth about it and still remember your answer: 'We can dream about rockets and the Moon until Hell freezes over. Unless the people understand it and the man who pays the bill is behind it, no dice. You worry about your damned calculations, and I'll talk to the people. Even if we continued our calculations we will not touch anybody. You may continue your theoretical studies; I will go public now, because this is where we have to sow our seed for space exploration.' You did and succeeded."[3] Von Braun gave his first public speech to the El Paso Rotary Club in 1947.

GENESIS OF A SPACE ADVOCATE

If we are to believe certain narrow-minded people—and what else can we call them?—humanity is confined to Earth from which there is no escape, condemned to vegetate on this globe, never able to venture into interplanetary space! That's not so! We are going to the Moon, we shall go to the planets, we shall travel to the stars, just as today we go from Liverpool to New York, easily, rapidly, surely; and the oceans of space will be crossed like the seas to the Moon! Distance is only a relative term, and ultimately it will be reduced to zero.[4]

— Jules Verne (1825-1905)

Ernst Stuhlinger and Frederick Ordway wrote a Wernher von Braun biographical memoir, *Wernher von Braun: Crusader for Space* (1996). Following are excerpts from the book: Von Braun's family tree can be traced back to the year 1285. His ancestors lived in Silesia, an eastern province of Germany along the upper portion on the Oder River. Among his ancestors were knights, landowners, governors, jurists, diplomats, generals, rectors of religious schools, but no engineers or scientists. Wernher's father, Baron Magnus von Braun, studied law and economics and chose the career of a government administrator. When Hitler came to power in 1933, Magnus von Braun resigned from government service and retired to his estate in Silesia. Wernher von Braun was born in 1912.

Asked about his early recollections of his son Wernher, von Braun's father answered with a twinkling in his eyes: "What I remember most vividly from the years of his childhood is the absolute futility of all my attempts to apply a bit of parental guidance to him. His growth rate was exorbitant, and I often thought that I should channel his outbursts of activity toward the more civil goals that were the accepted standards of society. Determination, fatherly strictness, diplomacy—nothing worked. Any attempt to admonish him, or to convince him of the inappropriateness of certain action, ran off not only like a drop of water, but like a drop of mercury, without leaving the faintest trace. I soon gave up—a reaction that really ran against my grain—and resigned myself to watching him grow, and to quietly paying the bills for broken windows, destroyed flower gardens, and other telltales that his early rocket experiments had left in the homes and yards of our neighbors. Looking back now over his life and mine, I am mighty glad that I took that stance; it was one of the best things I did in my life. As for his astounding talent for science and engineering, I can only play completely innocent. He certainly did not inherit this from me. I believe that he really got this from his mother, who possessed an exceptional accumulation of talents. In fact, I am convinced that my sons had the best mother they could ever have wished for."[5]

Wernher was confirmed into the Lutheran Church at the age of thirteen. His mother presented him with an astronomical telescope as his confirmation gift. This gift provided the spark that ignited his interest in the universe—our neighbors in space.

More than astronomy occupied young Wernher's mind, and no one knew this better than his mother. Years later, during the 1950s, Ernst Stuhlinger recalled her reminiscing about her son. "Learning always was easy for Wernher," she said. When he was ten, we decided to send him to the Französiches Gymnasium in Berlin to give him a somewhat broader basis of education. Around that time, I asked him once what he would like to do later in life. 'I want to help turn the wheel of progress,' was his answer. At times, he may have tried to turn a little too fast. Besides school, he involved himself in numerous projects with his friends. The boys were constantly on the move; they built all kinds of rockets, and they collected pieces of old automobiles from junkyards and built a new car which they tried out on vacant lots, with and without rocket propulsion. They certainly had a lot of fun, but as a side result, Wernher flunked in mathematics and physics. His father decided that our son really needed more guidance and control than he was willing to accept from his parents. So we enrolled him, in 1925, in the Ettersburg boarding school."[6]

Wernher would have been the last person to be disheartened. He read the book *The Rocket into Planetary Space* which Hermann Oberth, the great theorist of space flight, had published in 1923, and he immediately made two decisions: first to really learn mathematics; and second, to become a space pioneer. After studying the laws of Johannes Kepler he said, "Kepler's ellipses are for me what automobile races on the Avus speedway in Berlin are for others!"[7]

In 1927, while still at the Ettersburg School, he wrote a letter to Professor Hermann Oberth: "I know that you believe in the future of rockets. So do I. Hence, I take the liberty of sending you a brief paper on rockets that I wrote recently." Oberth responded, "Keep going young man! If you keep up your good work, you will certainly become a capable engineer."[8]

During the waning years of the nineteenth century and the opening decades of the twentieth, a new breed of dreamers and inventors began examining the feasibility of one of the most grandiose and exciting ideas in human history: manned travel beyond the Earth's atmosphere. The basis of their studies was the lowly rocket, the only propulsion system that their theoretical investigations revealed could operate in the near-vacuum of outer space. The time had arrived for an idea to merge into reality, as four outstanding men independently evolved sound concepts of how space travel could come about—Konstantin Tsiolokovsky (Russia), Robert Goddard (America), Robert Esnault-Pelterie (France), and Hermann Oberth (Germany).

Their imaginations were inspired by the great nineteenth-century writers of imaginative fiction, Jules Verne and H.G. Wells—men who made space travel sound exciting and, even more important, whose ideas sounded technically feasible to young boys with an aptitude for science and engineering.[9]

The fourth of the original pioneers of modern rocketry, Hermann Oberth, was the youngest, and the only one to witness the fruition of his early theoretical endeavors. At the age of seventy-five he personally watched the launching of Apollo 11 at the Kennedy Space Center, Cape Canaveral, Florida, on July 16 1969. Fate dictated that the theoretical work that Oberth had started, and the individuals inspired by his vision, would have far-reaching influence on the future.

More a theoretician, like Tsiolkovsky and Esnault-Pelterie, than a designer and builder, like Goddard, Oberth was born in 1894 in Transylvania, a German speaking region of the former Austro-Hungarian Empire. In 1923 his trail-blazing *Die Rakete zu den Planeträumen,* or *The Rocket Into Planetary Space,* appeared. Its preface contained four Oberth pronouncements that rang in the age of space travel:

1. With the present state of science and technology, it is possible to build machines that can rise to altitudes beyond the Earth's atmosphere.
2. With further improvement such machines can attain velocities that will enable them to stay unpowered in outer space without falling back to Earth. They may even be capable of leaving the Earth's gravitational field.
3. Such machines can be built so that people can ride them, in all likelihood without any harm to their health.
4. Under certain conditions the construction of such machines may become commercially profitable. Such conditions may come about within a few decades.

Oberth's book, with those ringing declarations, lit the fire of a running controversy in scientific circles that did not end until forty-six years later, when Neil Armstrong set his foot on the Moon.[10]

Societies comprised of enthusiasts for rockets and space flight sprang up in Germany (1927), the United States (1930), and England (1933). Von Braun joined the German Society at eighteen years of age. In 1932, when he was twenty, the German Army hired him for a program to develop rockets for defense, first in Kummersdorf near Berlin, then, from 1937 to 1945, in Peenemünde. The first long-range precision rocket-powered missile, the A-4 (V-2), was developed by the German Army under von Braun's technical direction at Peenemünde. During the war years, while he was working on the A-4 (V-2) guided missile project in Peenemünde, he had to stay quiet about the dream of space travel, except when he was among his inner-circle colleagues. When he expressed his dislike for the missile to be used against Germany's enemies, the SS put him in jail where he stayed for two weeks.[11]

In the meantime, while von Braun was working on rocket development for the German Army, there was another gentleman, in the Soviet Union, hard at work for the Russian Army: Sergei P. Korolev. He co-founded the Russian space enthusiast society, GIRD (Gruppa Isutchcniya Reaktivnovo Dvisheniya). Korolev was responsible for the development of the world's first intercontinental ballistic missile (ICBM), the R-7 in 1953. The R-7 launched the world's first satellite, *Sputnik I.* This event not only launched the satellite, it launched America's concern about the capability of the Soviet Union to attack the United States, or anywhere else in the world, using the ICBM/R-7. It wouldn't be long before Korolev and von Braun became adversaries (although they never met) during the next war—the Cold War.[12]

SURRENDER BUT NOT DEFEATED
Through the winter of 1944/45 the Allied forces were moving in on Germany from all sides. By the end of January 1945 the Soviet armies were less than 90 miles from Peenemünde ... they were deep into German territory. The distant rumbling of artillery fire was a constant companion in those days. An order was received to move all personnel and machines out—another order was received to defend Peenemünde against the Soviet army at all cost.

Von Braun would have none of it. With a number of his associates, he moved to Oberjoch, a small village high up in the mountains and took residence in Haus Ingeburg. It was there that they received the news of Hitler's suicide—the SS guards disappeared shortly after. On the morning of May 2, 1945, von Braun announced to his associates, "My brother, Magnus, who speaks English well, has just left by bicycle to establish contact with the American forces at Reutte. We cannot wait here forever ... "[13]

"Kommvorwärts mit die Hände hoch!" On hearing the command to put his hands up and come forward, the young German, Magnus von Braun, jumped from his bicycle and did as ordered. With his hands above his head, the German approached Private First Class Frederick Schneikert (Sheboygan, Wisconsin) and told him his name and that his brother, Wernher, had invented the V-2. Wernher and other members of the team were close by and wanted to be taken "to see Ike as soon

as possible." Schneikert turned Magnus over to First Lieutenant Charles Stewart, with the introduction, "This is Magnus von Braun. He claims to be the representative of over a hundred top German scientists located at a mountain inn. They want to surrender to us." Von Braun and his team were about to find a new home in the United States.[14]

THE CAMPAIGN FOR A SPACE PROGRAM

The first Germans arrived at Fort Bliss, Texas, in 1945 along with 300 freight cars full of V-2 components. A little over one year after the last V-2 was launched in Europe, the first was launched at White Sands Proving Ground (WSPG). Between April 1946 and September 1952, 67 V-2s were launched at WSPG. Two were launched from Florida and one from the aircraft carrier *Midway*. Of the 70, three fizzled on ignition, 20 reached altitudes of 2.5 and 65 miles (4 and 100km), and 47 soared to peak altitudes between 65 and 140 miles (100 and 213 km). They carried scientific payloads and were also studied to improve their guidance and control systems.

Billeted initially in Building H, Fort Bliss, a huge wooden airplane hanger to be used for assembling A4/V-2s and other missiles, the Germans wondered if they had been hasty in agreeing to come to America. The first days at White Sands and Fort Bliss were tedious ones for the Germans. Army policy prohibited social interaction between the officers and enlisted men and their former enemies. Any ideas the Germans had of advancing their work after leaving Peenemünde quickly vanished after they had been there for a while. Their mission was largely one of technical exploitation and assistance to others. Despite the humdrum chores of designing military rockets, the team maintained its old dream of space travel.

During their stay in Texas, they worked on missile projects and with their self-chosen study project, carried out jointly with members of the General Electric Company, of a V-2 booster, for a ramjet-driven cruise missile. Dr. Adolf Thiel reminded von Braun: "We walked the white sands while being told to learn something about ramjets. Not knowing much about them, we analyzed, designed, calculated, and dreamed about bigger rockets to the Moon. Not much response from our bosses!"[15] That project study also provided the framework for individual studies of guidance and control instruments, magnetic amplifiers, celestial mechanics problems, supersonic aerodynamics, electric propulsion systems, and other advanced work related to rocket flight. Von Braun, together with some of his associates, focused on a subject that had occupied his mind for more than twenty years, a manned expedition to Mars. He set his thoughts down in a book, *The Mars Project: a Technical Tale*.[16] The "Tale" describes, in great detail, an epic journey to the planet Mars. The manuscript is in two parts—the first is a fictional cast of characters traveling on a voyage to Mars; and the second is detailed with the proper mathematical proof which set the journey in motion.

When von Braun began his speaking tour and was writing articles for *Collier's* magazine and other publications promoting space travel, Thiel noted, "you started to get your thoughts into magazines. Some of us longhairs and purists didn't think very highly of that."[17]

Aside from giving many speeches, von Braun wrote popular articles for the general public. In particular were the popular, now historic, articles for *Collier's* magazine that appealed to *Collier's* staff members, artist Chesley Bonestell and author Cornelius Ryan. Bonestell recalled a luncheon in San Antonio, Texas (1951). "On a cold winter day, we had lunch, you as the guest of Connie Ryan and me (*Collier's* Magazine expense account). I asked, 'Dr. von Braun, had you ever thought of going to the Moon?' 'Call me Wernher. Yes, indeed I have,' starting to draw a rocket on the paper napkin. 'What, not streamlined? Do you want it streamlined?' 'No, you design it the way it should be.' The craning necks of the rival magazine editors at the table from *Look* and *Saturday Evening Post* caused me to say, 'well, never mind at present—we'll go into it later.' And we did indeed, next day at lunch at a different hotel. After lunch you and Cornelius talked about a possible *Collier's* contract."[18] Von Braun had convinced a highly skeptical Ryan that space travel was possible and desirable. The fact that he succeeded is evidenced by the subsequent articles in *Collier's* magazine.[19]

In 1950 the team was transferred to Redstone Arsenal in northern Alabama, near the town of Huntsville, where they carried on their work. In 1960 they were administratively transferred to the newly organized civilian space agency, NASA.

Their activities were acknowledged as a true milestone in the evolution of space flight by the professionals, but von Braun realized that unless "the man who pays the bills," the taxpayers, understand the scope and value of such a project, the effort to achieve the goal of manned missions into space would not be achieved. A change in strategy would be necessary. Von Braun decided to make the change; in a dramatic outburst he set out his plan to "speak to the man who pays the bills—the tax payer."

*** END ***

— Chapter 2 —

The First Public Speech—Many to Follow

The El Paso, Texas, Rotary Club invited von Braun to speak. After the Army gave its approval, on 16 January 1947, von Braun gave his first speech to a public audience in the United States, the El Paso Rotary Club. This would be the first time he was entirely free to express his own thoughts about the evolution of the rocket from an Earth-bound weapon to a vehicle that would one day carry human astronauts to the Moon and to planetary space.[20]

Future Development of the Rocket [21]

GENTLEMEN: WHEN, ABOUT 14 DAYS AGO, I RECEIVED THE HONORABLE INVITATION to talk to you on rockets, I thought it might be perhaps a good idea to tell you something about the future prospects of rocketry. At the present time there is so much being published on future possibilities of rockets and traveling through space, that it might be of particular interest to hear what of these things can be called serious and what nonsense. Since I just learned that Colonel Turner of White Sands talked to you a few weeks ago on the rocket tests carried out at White Sands, you will probably be familiar with definitions used in general rocketry discussion, so that I may go right on with the subject.

The principle of a rocket is very simple. The main difference between rocket propulsion and any other propulsion is that the rocket does not need any air for propulsion. The rocket gets its propulsion from the expulsion of the gases which are burned in the rocket motor itself. A loaded rifle pushes with the same force at the back side of the rifle as it expels the bullet. This is the backstroke you feel when you fire a rifle, and this same reactive force is used in a machine gun to repeat and reload the barrel. If you put a machine gun on a little car, this car will certainly begin to run backwards. This is the principle of the rocket. The rocket motor shoots millions and millions of molecules out of a nozzle, the nozzle being the barrel and the molecules being the bullets of our machine gun. Like a machine gun, the rocket does not need air. This offers unique possibilities of traveling in empty space.

Now the rocket is being built in principle. Two tanks—one for liquid oxygen and one for alcohol. Both are pumped into a combustion chamber where the combustion takes place. The combustion pressure shoots the exhaust gases out of the nozzle.

You might ask me why the rocket was used first in a larger scale as a weapon of war. I can't explain. It seems to be a law of nature that all novel technical inventions that have a future for civilian use start as weapons—e.g. we have the same development with atomic energy being used first as a weapon. All of these things bring unique chances to improve the standard of living. The only hope we can have right now is that in the future an everlasting peace will allow the use of the rocket in its peaceful application.

In principle, as we enlarge the V-2 rocket, we have a sort of "space ship." We could put people in and then have people climb up outside the atmosphere. There is no limit to increasing the size of the rocket, just as limits are being ignored in ships and airplanes being developed.

We can't put man in the V-2 at White Sands now, as it is difficult to get a man down again alive. Many offers from volunteers have been given, but the problem of landing him is difficult due to the fact that on the way back the missile picks up a tremendous speed, and even the parachute which he would use would burn up. It is necessary to provide other aerodynamic means of getting him down again.

The disadvantage of an increased V-2 rocket, which merely provides space enough to hold a crew, is that the performance would not be much more that the V-2 rocket itself.

This illustration shows a rocket that would be able to leave the gravitational field of the Earth. This missile can reach a velocity so high that the upper stage of the rocket would not fall back on the Earth—it would reach orbit velocity ... [it] could escape the gravitational pull of the Earth. Total size of such a rocket is about 90 feet—approximately the height of a 6-floor building. No reason for size being a hindrance, as ships and airplanes are being built such sizes.

When a rocket has reached its final velocity and fuel is exhausted and propulsion stops, no gravity is in the rocket any more. It is not true that entire lack of gravity inside the rocket occurs only outside of the Earth's gravitational field or at the famous "neutral point" between Earth and Moon. Any gravitational sensation inside the rocket vanishes once the power is shut off and the rocket coasts without propulsion, for there are no forces between the men and the floor or instruments in the rocket, because all go at the same speed.

When the "cut-off" of the rocket takes place, there is a queer sensation in the cockpit of the rocket—and all will just float and fall. The gravitational sensation returns when the rocket reenters the atmosphere and is slowed down by air drag, and because of this deceleration the man feels then that he is falling forward. The lack of gravity is not critical. It has been proved that for the human body gravitation is not necessary. Man can eat and drink, etc., lying down and even standing on his head. Thus, there is no great problem or principle difficulty to get human beings up to the empty space. The problem is getting them down again. It is difficult to bring even the instruments down since at reentering the atmosphere the speed is so high. However, if we tilt the rocket in flight on a shallow or horizontal plane, we have the high horizontal speed and it is possible to glide. We used this way of increasing the range of the V-2 in Germany.

The A-9 rocket was built in Germany in 1943 and 1944. It is a V-2 with wings, and the flight is stretched by means of these wings. Two successful flights were made and the range was found to be about twice that of the trajectory of the V-2. This method offers the possibility of landing the rocket. Even at very high speeds, such as occur when a space ship reenters the atmosphere in a tangential path, these wings can keep the rocket flying like a plane. They allow the speed to slow down and go deeper and deeper, and finally land like a plane.

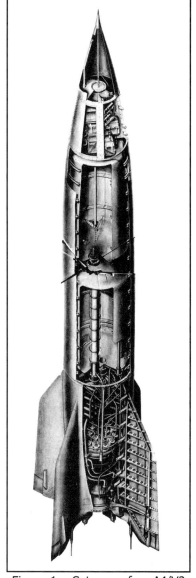

Figure 1 – Cutaway of an A4/V2

This same problem—putting wings on rockets—is being approached by airplane builders. This design looks very sound and well-prepared scientifically, and certainly offers all chances of reaching the goal. If the Bell Company successfully manages to fly this plane in the supersonic speed area, the most ticklish problem, namely, the passing of the transonic speed area with manned vehicles, would be solved. There are no further principle aerodynamical difficulties to be expected after the problem is licked.

The following may seem fantastic to you. The illustration shows the Earth and an orbit around the Earth from which the rocket would not fall down.

Make up a speed of 22,000 feet per second—four times that of the V-2—then the upper stage of the rocket keeps on coasting around the Earth without further propulsion. This speed offers certain possibilities: to set up a platform in the empty space, and keep it manned and used for

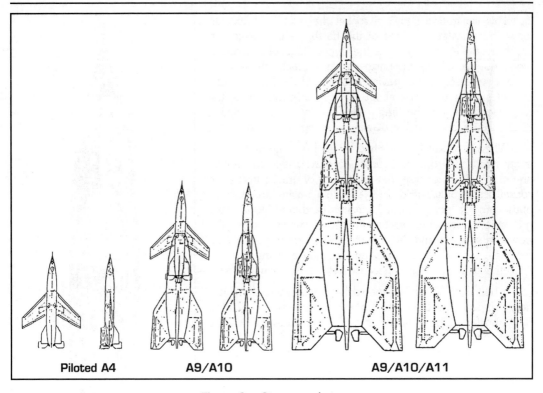

Figure 2 – Stages rockets

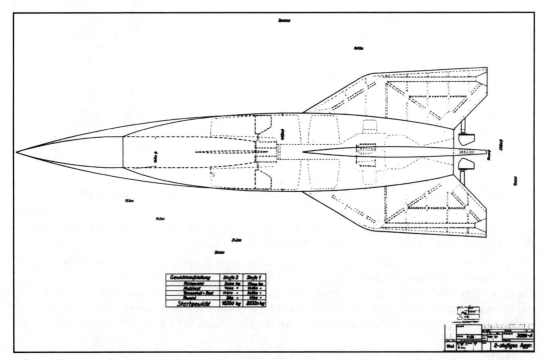

Figure 3 – A-9 rocket

Figure 4 – Bell-X-1

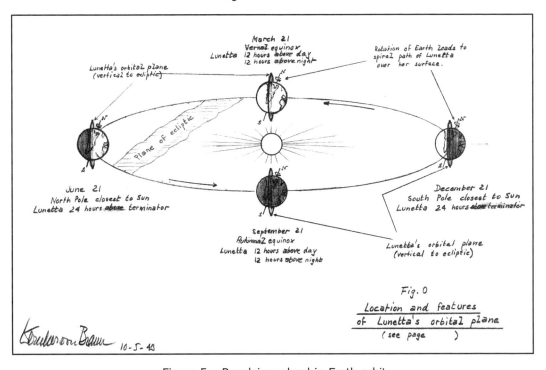

Figure 5 – Revolving wheel in Earth orbit

various purposes ... offers new possibilities in efforts to explore the universe. Such a platform would look like a revolving wheel.

The first problem in this platform is to provide normal gravity conditions for people up there, this being done by shaping the platform as a slowly revolving wheel. When this wheel has a diameter of 150 feet and makes one revolution in 11 seconds, the centrifugal force provides an acceleration of exactly 1 g—the normal gravitational acceleration on Earth. Secondly, the wheel has to be "pressure conditioned"; this is done according to the same principle as in use in submarines, stratosphere planes, etc. Thirdly, temperature control has to be developed. This is simpler than it might seem, for the far-spread belief that it is terribly cold in the universe is not true. Temperature is the speed of the movement of molecules; thus, where there are no molecules there can be no temperature either. A sphere, suspended in the universe, not too far away from Earth, acquires a temperature of plus 16° C, which, by the way, is also the average temperature of the Earth (averaged over all climates and seasons). This temperature represents the equilibrium at which the absorbed heat of the Sun rays equals the heat losses due to radiation of the sphere itself. If we make the sphere black on the Sun side and blank at the shadow side, the absorption of the sphere will go up to plus 200° to 250°C.

Then there is the need for electric power in the station. Here the platform is equipped with a large mirror which concentrates sunrays on a steam boiler. The steam drives a turbine, which in turn drives a generator, [with] the cold steam being condensed to water in a "surface radiation condenser."

How to build this installation into empty space? It can be built of rubber-like material so that it can be stored in reasonably-sized rocket cargo rooms. After unloading it can be inflated by inner pressure like a balloon. This platform could be used for the future development of interplanetary travel. If the rocket refuels at the station and accelerates in the same direction in which the station flies with 22,000 feet per second, until the rocket makes 33,000 feet per second, it will enter a long-stretched Keplerian ellipse, the remotest portion of which leads beyond the orbit of the Moon. After 70 hours of coasting, the rocket would finally be in a position on the backside of the Moon, which man does not know about, and from which photos could be taken.

According to Kepler's Laws of the planetary orbits, the rocket would then coast back with increasing speed and approach the orbit of the platform again—however, with a speed that would be about 8,000 feet per second faster than the platform's. With a short period of counter blast of the rocket motor, the rocket's 30,000 feet per second would then be decelerated to the platform's 22,000, so that the rocket would land at the platform again. In this way the platform is an ideal place to start trips to the Moon. The platform could thus be used as a permanent "harbor" of space ships, and men could land back by gliders that wouldn't need fuel, leaving the space ships "on roadstead" [harbored] in the neighborhood of the platform.

What practical reasons exist for the achievement of these developments? There is, first, the high-speed traffic from continent to continent. Europe could be reached in three-quarters [of an] hour—making [the] economic bands of Earth tighter.

Interplanetary travel—No man can say exactly what it will bring. The first man who puts his foot on the Moon, or another planet, would be much in the position of Columbus when he discovered the New World. With mankind exploring and conquering other planets, her future history is unlimited and unpredictable.

*** END ***

– Chapter 3 –

1945-1951

On August 27, 1949, the following invitation was sent to Professor Wernher von Braun in Fort Bliss, Texas, from L.J. Carter, Secretary of the British Interplanetary Society: "At a recent meeting of the Council of this Society, it was proposed and unanimously agreed that I be instructed to extend to you, in recognition of your great pioneering activities in the field of rocket engineering, the invitation to become an Honorary Fellow of this Society." Von Braun humbly replied, "I consider your invitation an outstanding honor despite the grief of me and my associates brought to the British People."

In 1951 von Braun was invited to present a paper at the Second Astronautical Congress meeting in London. He agreed to write the paper, but notified Dr. R.L. Shepard of the British Interplanetary Society (BIS) that he would not be present to read it. Val Cleaver of the BIS stated he'd be proud to read the paper in his place. C.W. Chillson, Program Chairman of the American Rocket Society (ARS) suggested to von Braun that the paper should be submitted under the auspices of the ARS. Von Braun agreed to comply with the ARS request and wrote Cleaver of his decision. Cleaver responded to von Braun: "It will give an appropriately cosmopolitan flavor to our international meeting if the opening paper is by an author of German birth (I don't know whether you are now a U.S. citizen?), who is an Honorary Fellow of a British Society [BIS] presenting his work under the auspices of an American one—regardless of whether an Englishman or an American reads it."[22] British rocket engineer Val Cleaver brought to mind events surrounding his experiences during World War II. "I tried to help convince our authorities that the Intelligence reports about your V2 from Peenemünde might be true. One evening in 1944, a double bang over London (with no prior air raid warning) told me I had been right. Within months, the war was over, and we were proud to make you one of the earliest Honorary Fellows of our British Interplanetary Society."[23]

The paper was read by Frederick C. Durant III, at that time one of the directors of the American Rocket Society (ARS). "I learned that von Braun had a paper ... but he was not allowed out of the country for security reasons...," Durant remembered, "So I sent him a letter ... suggesting that we meet ... we met in New York, and within 3 to 5 minutes we felt closely associated." Durant presented the following paper in von Braun's place.[24]

THE IMPORTANCE OF SATELLITE VEHICLES IN INTERPLANETARY FLIGHT [25]

THE GIANT AIRLINES WHICH FLY THE TRANSCONTINENTAL AND TRANSOCEANIC lanes are the result of several decades of persistent effort to improve aerodynamic configurations, mass ratios and performances of power plants. But any antiquated small airplane can fly around the globe if we subdivide her flight into a number of short hops and allow her to refill after each landing.

The trouble with rocket flights into the planetary world is that interplanetary filling stations are few and far between. Therefore, a Mars- or Venus-bound rocket ship, taking off from the Earth's surface, would have to carry all propellants required, down to those for the final power maneuvers of the return voyage. As a result, it would have to be an immensely large and heavy ship with a power plant of truly astronomical proportions.

If we want to avoid the knotty problem of developing such a super space ship, we should first build those filling stations. It is one of the main tasks of the satellite vehicle to establish such a filling station for interplanetary flights.

There is a widely held belief, even among space travel enthusiasts, that interplanetary voyages with chemically-propelled rocket ships involve such tremendous quantities of propellants that the technical feasibility of such trips must be questioned, even for very limited payloads. It has therefore become a habit to hint at mysterious "atomic propellants and nuclear rocket drives" which could do the trick better. I have no desire nor intention of decrying the eventual application of this source of power to navigation of interstellar space. When referring to technological advances, the word "impossible" must be used, if at all, with utmost caution. We should not forget, however, that all speculation on nuclear power sources for rockets is still founded on rather shaky ground. Thus, we render our cause a service of dubious value by referring to such vague potentialities. I, for one, am not yet convinced that within the next quarter of a century we will have a nuclear rocket drive that could economically compete with chemical power plants.

Factors such as required thrust ratings or radioactive contamination of launching sites render the problem of developing a nuclear rocket ship suitable for takeoffs from Earth much more involved than the development of an orbit-to-orbit space ship for exclusive use in empty space. Therefore, the first atomic-driven rocket will probably start from the orbit rather than from the Earth's surface. The main burden of any voyage into the planetary world, viz., the prefatory supply flights into an orbit around the Earth, would even then have to be borne by chemically-propelled ships.

Rocket designers generally agree that at least three stages are required for a satellite rocket ship with chemical propellants. If such a ship is to serve for the preparation of an interplanetary flight originating in the satellite orbit, it must be capable of carrying aloft a liberal amount of payload.

The ship's last stage must be able to safely return to the Earth's surface after unloading its payload in orbit. Such a return is possible if we equip the upper stage with wings. By a short power maneuver in the orbit, the circular velocity would be reduced by an amount adequate to induce the upper stage into a "landing ellipse" whose perigee is located in the tenuous upper layers of the atmosphere. In a long-stretched glide, first flown with a negative, later with a positive angle-of-attack, it would gradually be decelerated by air drag, until it descended to subsonic speed and subsequently landed on a tricycle landing gear in airplane fashion.

Optimum investigations indicate that a given payload can be carried into the orbit with less takeoff weight if the number of stages of the satellite vehicle exceeds the minimum of three. The exact figure for the optimum number depends upon the choice of propellants and the mass ratios considered realistic for the individual stages. Most authors believe that an optimum satellite vehicle should have four or five stages, and further refinements such as jettisoning of power plants and tanks have been suggested.

However, if we think of using the satellite vehicle for the preparation of an interplanetary voyage originating in the orbit, we must bear in mind that a great number of supply flights to the orbit will be required. Inasmuch as the last stage can glide back to Earth, it may be used for repeated flights. It would greatly facilitate the supply operation if the booster stages could likewise be salvaged and repeatedly used. Careful studies of this problem have convinced me that such booster recovery is definitely possible. If an expedient track of ascent is chosen, the burnt-out boosters can be decelerated by means of steel mesh brake parachutes in such manner that the terminal rate of descent does not exceed 150 to 200 feet per second. Since empty boosters have enough buoyancy to float, they should be landed in the ocean, and their rate of descent at the moment of ditching could be reduced to virtually zero if the parachute descent were decelerated by means of powder rockets ignited by a proximity fuse immediately before impact. In this manner, the boosters would very gently take to the water and not suffer any damage. It would appear possible either to take the floating boosters aboard a salvage boat or to tow them back to the launching site.

The greater the number of stages of the satellite vehicle, the more involved this booster recovery problem will become. Not only that more boosters must be salvaged after each flight, but the last booster will land so far from the takeoff site that its shipment back to the base will delay the entire supply operations to the orbit. Therefore, the question as to how many stages should be chosen for the satellite vehicle should not be investigated exclusively from the aspect of minimum takeoff weight for the individual ship, but also in the light of the master plan for a supply operation such as is required for the preparation of an interplanetary voyage starting from the satellite orbit.

Let me return to the winged descent of the satellite vehicle's last stage. This descent raises a number of interesting questions.

What altitude should be chosen for the perigee of the landing ellipse, in order to convert the elliptical approach into a glide path? The air density at this altitude should be sufficient for generating enough downward lift to prevent the last stage from receding from the Earth again on the second leg of the ellipse, and yet it should be small enough to avoid excessive skin temperatures during the initial phase of the glide through the atmosphere. I found that an altitude of the perigee in the order of 50 miles is expedient.

What skin temperatures will be encountered during this rapid glide? Despite the high altitudes at which the glide is performed, the heat influx into the ship's skin is very great. The only reasonable way of controlling skin temperatures appears to be the use of skin material that may safely be used at temperatures where re-radiation equals heat influx. Inasmuch as, according to Stefan-Boltymann's law, the rate of skin radiation increases with the fourth power of the absolute temperature, skin temperatures well above 500° C must be allowed. It was found that the entire glide of the third stage from the satellite orbit can be performed in such manner that skin temperatures never exceed 730° C. Heat resistant steels are available, which can handle such temperatures.

One can always reduce the skin temperatures during the glide by applying a smaller wing-loading, i.e., by equipping the last stage of the satellite vehicle with larger wings. These enable the last stage to perform [a] critical portion of the glide at higher altitudes, where the heat influx is reduced on account of the lesser atmospheric density. The choice of the wing-loading is also important in the matter of landing speed, in that the latter decreases with decreasing wing-loading. If we choose a wing-loading whereby the skin temperature during the glide does not exceed 730° C, the last stage of our satellite vehicle would have a landing speed of about 65 miles per hour, less than the touchdown speed for a conventional airliner.

For such a wing-loading the length of the glide-path from the perigee of the landing ellipse to touchdown is about 14,000 miles, which is a little more than 50 percent of the circumference of the Earth. Inasmuch as the preceding free-coasting flight through the landing ellipse also leads half around the Earth, the last stage performs one full circumnavigation from the deceleration maneuver in the orbit to touchdown. The flight time from orbit to landing would be roughly two hours.

Let us now suppose we had a reliable satellite vehicle. Would it really render possible the exploration of the other planets?

During the past three years I have spent my spare hours making a comprehensive study of an expedition to Mars, which is at present in the process of publication.

The study deals with an expedition of 70 men, traveling in ten space ships from a satellite orbit around the Earth into a satellite orbit around Mars. Landing on Mars' surface is performed with three winged "landing boats," two of which are capable of returning to the space ships left circling in the circum-Martian orbit. Only seven of the ten space ships, with the original crew of 70, return to Earth. The expedition is prepared with the help of a fleet of 46 three-staged satellite vehicles. In 950 flights, these satellite rockets haul space ship components, propellants, landing boats, auxiliary equipment and crews to the "orbit of departure" where the space ships are assembled and whence the Mars fleet takes off.

The entire investigation is based on chemical propellants. In order to avoid the logistic problem involved in the use of large quantities of liquefied gases, the calculations have been based on hydrazine (N_2H_4) for fuel and nitric acid (HNO_3) for oxidizer.

The three-stage satellite vehicles are the largest units required for the entire undertaking. They would be capable of delivering a payload of about 39 metric tons to the orbit in each of their flights. The overall length of the satellite ship would be about 200 feet, the maximum diameter 65 feet and the takeoff weight 6400 metric tons. Both boosters can be recovered by combined parachute and rocket landing mechanisms, whereas the third stage would glide back with the help of its wings.

The design of the orbit-to-orbit ships can greatly benefit from the fact that they operate exclusively in empty space. They require no streamlining and not even a hull. In order to facilitate their hauling into orbit, their spacious tanks could be made of collapsible fabric or plastic. These tanks could be freely suspended in an open lightweight structure. The stage principle, so essential for the satellite vehicles, could be abandoned. Instead, the empty propellant containers would simply be separated from the open ship structure after each power maneuver. Inasmuch as the ships start their voyage from an orbit, in which weight is completely balanced by centrifugal force, the thrust rating of their power plants may be limited to a fraction of the ship's initial weight. The entire round trip to the circum-Martian orbit could be conducted with a rocket power plant whose thrust rating does not exceed that of the satellite vehicle's last stage, and this same power plant may be used from the initial maneuver of departure down to the terminal maneuver of return to the orbit around the Earth. It is evident, then, that all these factors greatly simplify the task of designing interplanetary rocket ships. We should remember, however, that such possibilities can only be fully exploited by a strict separation of the tanks of the interplanetary ships from those of the satellite vehicles, which must be designed for flights through the atmosphere and for ascent within the gravitational field of the Earth. Indeed, the design of an interplanetary orbit-to-orbit ship appears to be much less of a task than the development of a satellite vehicle. We may say, therefore, that the satellite vehicle is not only the first, but the decisive step toward interplanetary flight.

A flight to a circum-Martian orbit would not be complete if it were not climaxed by a descent to the surface of the Red Planet itself. The orbit-to-orbit ships, solely designed for use in empty space, are unsuited to this task. Our 70-man expedition needs special landing boats, which could descend from the circum-Martian orbit in very much the same manner as the last stage of the satellite vehicle returns from its orbit to the surface of the Earth. The design of such Mars Landing Boats is complicated by a number of factors, however.

1. Whereas the last stage of a satellite vehicle would land with empty propellant containers, the landing boats would have to carry enough propellants down to Mars' surface to enable them to return to the orbit.
2. The bottom density of the Martian atmosphere is only about one-twelfth of the sea-level density of the terrestrian atmosphere, and the lift produced by a wing of equal size and at equal landing speed would be correspondingly smaller. On the other hand, the gravity on Mars amounts to only 38 percent of that on Earth, and the lift actually required is reduced to that extent. In taking both these factors into account, it turns out that for the same landing speed, the wing area of a ship of a given mass landing on Mars must be four times as large as for landing on Earth. In view of the before mentioned fact that a Mars Landing Boat must descend heavily loaded with propellants, it is obvious that it requires very large wings. Prior to the return of the boat into the circum-Martian orbit, these wings can be shed, of course. Considering Mars' weak gravitational field, the re-ascent can easily be performed with the one-stage rocket constituted by the boat's hull.

If we abandon one boat on Mars' surface and return the landing party to the orbit in the two remaining boats, we can fill more than a hundred tons of payload into the first boat, because it requires no propellants for the re-ascent. In this manner we can carry down to Mars' surface a load adequate to supply a landing party of about fifty people for more than a year.

How can we haul these heavy landing boats from Earth into the circum-Martian orbit? Fully loaded with propellants, each of these boats would weigh about 200 tons.

Any space ship designed for the round trip between Earth and Mars must arrive in the circum-Martian orbit with a sufficient propellant supply left for the return voyage. The ones I investigated would still weigh a little over 400 tons after settling in that orbit. Therefore, all we have to do is abandon those three orbit-to-orbit ships that carried the landing boats, after arrival in the Mars orbit, and perform the return voyage with the seven remaining craft. Over and above the landing boats, each of these one-way orbit-to-orbit ships could then haul about 200 tons of additional payload into the circum-Martian orbit.

Difficulties may arise in shipping a landing boat with its airplane hull piecemeal in the crowded cargo holds of the satellite vehicles, and in assembling it in the orbit of departure. It would, therefore, appear expedient to have the landing boat fly to the orbit of departure under its own power. It will entirely be capable of doing this if we mount it on two standard boosters of the satellite vehicle, in lieu of the normal third stage.

It would exceed the scope of this paper to present you with a complete proof that all this is a real and serious possibility. For this proof I must refer to the publication mentioned before.

You may be interested in a tabular compilation of the results, however. For this reason I prepared four tables.

Table 1 shows the main data of the three-stage satellite vehicles.

TABLE 1 - THREE-STAGE SATELLITE VEHICLE

First Stage:	Thrust	12,800 tons
	Burning Time	84 sec
	Velocity at cut-off	2,350 m/sec
Second Stage:	Thrust	1,600 tons
	Burning time	124 sec
	Velocity at cut-off	6,420 m/sec
Third Stage:	Thrust	200 tons
	Burning time (ascent)	73 sec
	Velocity at cut-off	8,260 m/sec
Total ship:	Length	200 ft
	Diameter	65 ft
	Take-off weight	6,400 tons
	Propellant supply	5,583 tons
	Payload (hauled to orbit)	39.4 tons
	Landing speed (third stage)	65 mi/h

Principle data of the orbit-to-orbit ships are compiled in Table 2. The figures apply to the seven ships designed for the round trip involving the following four power maneuvers:

1. Departure from orbit around the Earth.
2. Induction into orbit around Mars.
3. Departure from orbit around Mars.
4. Induction into orbit around Earth.

TABLE 2 - ORBIT-TO-ORBIT SHIP

Thrust	200 tons
Total propellant supply	3662.5 tons
Total length	134 ft
Diameter	95 ft
Maneuver 1	
Initial weight	3720 tons
Terminal weight	906 tons
Burning time	3965 sec
Maneuver 2	
Initial weight	902 tons
Terminal weight	410 tons
Burning time	658 sec
Maneuver 3	
Initial weight	408 tons
Terminal weight	186 tons
Burning time	298 sec
Maneuver 4	
Initial weight	185 tons
Terminal weight	50.5 tons
Burning time	163 sec

Table 3 shows the principle data of the Mars Landing Boats. The two boats designed for landing and re-ascent can each carry a payload of 12 tons down to Mars and 5 tons back to the circum-Martian orbit. The third boat, to be abandoned on Mars' surface, is capable of landing with a payload of 125 tons. This gives us a total payload of 149 tons available for the landing party. Aside from such vital necessities as food, oxygen and water, such payload may include ground vehicles suitable for land excursions on the Martian surface, inflatable rubber houses wherein the landing party can live without pressure suits, and similar luxuries. Luxury incidentally, may be the wrong word, for you will agree that only with such elaborate facilities an extended stay and a successful exploration of a strange planet will be possible.

Table 3 – Mars Landing Boat

Thrust	200 tons
Initial weight	200 tons
Landing weight	185 tons
Landing payload	
(one-way boat)	125 tons
(two other boats)	12 tons
Landing speed	20 mi/h
Take-off weight (re-ascent)	138 tons
Payload for re-ascent	5 tons
Burning time (re-ascent)	147 sec
Velocity at cut-off (re-ascent)	3,700 m/sec

Table 4, finally, compares the scope of such an expedition to Mars with the expenditures involved.

Table 4 – Relation of Time and Weights of Propellants of an Expedition to the Planet Mars

Number of participants	70
Total duration of voyage	2 years 239 days
Traveling time—Mars	260 days
"Waiting time" in circum–Martian orbit	449 days
Hereof, stay of landing party	ca. 50 men
on Mars' surface	ca. 400 days
Traveling time—Mars	260 days
Total payload available in circum–Martian orbit	ca. 600 tons
Total payload available on Mars' surface	149 tons
Number of orbit-to-orbit ships	10
Number of Mars Landing Boats	3
Required number of Satellite vehicles	ca. 46
Required number of supply flights	950
Time required for supply operation	ca. 8 months
Propellants required for supply operation	5,320,000 tons
Propellants required for interplanetary voyage	36,600 tons

At first glance, 5 million and 320 thousand tons of propellants appear to be a startling figure indeed. But let us take a second look. This is the cargo of 443 tankers with a capacity of 12,000 tons each. According to the official statistics, the Berlin Airlift consumed about one-tenth of this amount in high-octane aviation gasoline! And all this just because of a little misunderstanding between diplomats! Compared with the fuel consumption of wars, the requirements for an interplanetary expedition would indeed pale into insignificance.

Let us hope, therefore, that by the time mankind is ready to enter the "cosmic age," wars will be a thing of the past, and, instead of paying taxes for armament, people will be ready to foot the fuel bill for a voyage to our neighbors in space.

*** END ***

English rocket engineer, Val Cleaver wrote von Braun: "[Commander Fred] Durant did an excellent job of presenting your paper, which created a lot of interest and attracted more press comment than any other, which is saying a lot."[26]

– Chapter 4 –

Celebrating the fiftieth anniversary of the first International Airshow in 1959, the City of Frankfurt on Main, Germany, invited von Braun to give a keynote address with the title *"The Beginning of Space Flight."* He began by reminding the audience of the haunting dilemma into which almost any kind of creative work can lead: the same machine that can bring uncounted benefits to mankind may also be used for destruction and violent death. He said, "We must not overlook the depressing although obvious fact that the same modern rockets that have opened the doors to the scientific exploration of space can also be equipped with atomic warheads and used as military weapons of horribly destructive power. There are simply no valid rules that can guide our decisions in this heartbreaking conflict, which is a bitter and tragic reality for all those who are entangled in such situations. Should Dr. Einstein, when he wrote down his famous formula for the relation between energy and matter, have dropped his pencil in despair because he had the vision of the release of unthinkably large amounts of atomic energy? Should Otto Lilienthal have discontinued his heroic glider flights because the possibility of a military misuse of the still unborn motorized airplane dawned on him? And should we rocket builders of today stop our efforts to open the universe to human exploration because rockets, just like airplanes, can be used for military purposes? It is just unfair to ask us scientists and engineers such questions to which even the wisest statesman and church leaders do not find answers."[27]

Val Cleaver and von Braun carried on extensive, painstaking debate through technical letter writing in the years 1946 until Cleaver's death in the 1970s. One of the their soul-searching issues was on the question of weapons as possible deterrents of war. Cleaver's argument: "Since there have always been weapons in the past, the deterrent potentialities of weapons are disproved." Von Braun: "The ethics of the Western World alone, or the general public desire for peace, just is not enough to keep the Communist tide in check. It is 'The Bomb in Being' that does the trick."[28]

In an attempt to contribute to the missile development programs, von Braun presented a proposal to Army Ordnance in 1949 suggesting the development of a large multipurpose booster rocket. This booster could be applied as an initial thrust stage to all the missiles under consideration, greatly enhancing their performance and usefulness. Shortly after he completed the report, he wrote the following speech/report. It was at the time when the Soviets had just exploded their first atomic bomb, and when they were building up their military power at a terrifying pace. It was a time of growing anxiety; the dream of eternal peace that had blissfully spread over many minds began to vanish quickly: "The West finds itself obliged to arm to the teeth in the interest of maintaining an uneasy peace in this tortured world ... first priority must be given to the deterrent power."[29]

Fred Durant read the paper for von Braun at an IAF meeting in Zurich at the Fourth International Astronautical Federation Congress, August 1953.

WE NEED A COORDINATED SPACE PROGRAM! [30]

How far in the future is the flight of man through space?

There was a heated debate on this subject last year. It was echoed and re-echoed by the press on both sides of the Atlantic in considerable detail. We may all be well satisfied that most of the commentaries cast a little or no doubt upon the ultimate feasibility of the project. Their questions were: "When? and How?" rather than, "Can it be done at all?"

On the other hand, it somewhat unfortunately began to appear that there existed two schools of thought among the proponents of space navigation, an appearance which I believe to be fundamentally in error, despite its reiteration in a series of articles. According to these articles, the adherents of one school advocate extreme caution and a slow, step-by step, almost organic developmental process (thus bringing us very gradually closer to space travel), while the other school wishes to rush in where angels presumably might fear to tread and to construct out of hand one or more multi-stage, manned space vessels. Since I myself have been variously characterized as belonging to the latter group, I am particularly happy to be afforded the opportunity to elaborate my views before this distinguished gathering, and with special reference to this important question. In point of fact, I am in considerable doubt whether the really debatable issue has yet been clearly defined.

No one with any experience whatsoever in the development of large rockets—or what presently are known as large rockets—could possibly seriously maintain that we are in a position today to leap from the status quo to manned rocket ships without a series of intermediate steps. Actually, it is no short leap at all, but rather an extended journey. Do not forget that any long journey requires meticulous preparation, even into well-explored regions. This kind of journey requires that one take careful account of such matters as the most desirable route to be followed, attainable stopping off places, and the like. When there is a whole ocean to be crossed, a prudent navigator will sit down over his charts and lay out the course he proposes to follow, and only thereafter will he set his sail. Let there be no mistaking that I advocate an entirely parallel procedure with respect to our coming voyage into space! Nor do I advocate a coordinated space program because I believe that the path into the void is short and easy; on the contrary, I am convinced that the road is long and hard—much too far for us even to think of rashly sailing out into the blue which we know soon becomes inky black to the spacefarer!

Now, the "cautions school" contends that it is too early for us even to sit down over our charts; and this is, I think, the moot question of the whole debate. They seem to hold that the seafarer on the oceans of space might best venture outside the harbor for a stretch before laying off the rest of the course. How can he, claim they, maintain a predetermined course in the face of unknown obstacles, cross-winds and cross-currents?

Let us, for a moment analyze the nature of these obstacles and deterrents, the first of which is whether our available propellants are sufficient to achieve for us the initial and primary important step of orbital flight. Can we get sufficient performance from any propellant combination? Long-range rocketry and spacefaring problems differ from many other technical questions—and may I say that it is a fortunate difference—in that the required performance of a rocket may be clearly formulated in terms of terminal velocity. We need not fear that the laws of mechanics will play us false in this matter, and so we are in a position accurately to compute the dimensions and weight of a rocket intended to fly faster and carry greater payloads than its smaller predecessors. The data derived from the latter have been proved accurate. We are familiar with the multi-stage principle and we know how to utilize it economically in order to augment terminal velocities. We are free to be conservative as to our assumptions. We have data derived from well-tested, existing propulsion units, and these data will serve as a basis for computations relative to new ones. Our methods of computing stresses in rocket bodies are tried and true, so that we can design the latter without fear of failure. Thus there is not the slightest reason for anxiety concerning our performance predictions. Our calculations definitely show that we can shoot a rocket into a satellite orbit using propellant combinations which are not only available, but which have been thoroughly tried on the test-stand! If we design for a modest payload, a satellite rocket will remain relatively modest in size. If, on the other hand, our payload demands are more exacting, the rocket will naturally become considerably larger; and if we contemplate a manned rocket—such as might deliver parts for a manned satellite to the orbit of the latter—the dimensions and weight will be very large indeed. This is particularly true in the light of the fact that the upper stage of the manned rocket must return safely to Earth. But even with this more ambitious project we obviously do not violate any mathematical laws of limitations.

Should we, perhaps, be in violation of a certain other law of limitations, namely that of our technological capacity? Technical history abounds with incidents which offer the most convincing proof that scaling up a normally healthy principle does not necessarily lead to difficulties which are in themselves basic. Outstanding examples are the Eiffel Tower, the Golden Gate Bridge and the Zeppelin Airship. Every man of those who helped develop the A-4 rocket,[31] including myself, was considerably abashed at its size, and we used to gaze up at the ogive of its lofty nose and wonder whether the seemingly monstrous prototype could really do what the calculations said it would—namely thrust itself upward and attain a velocity some five times that of sound. In those days we really faced a barrier of unanswered questions! No steerable body had ever acceleratively forced through the sonic region, and some of our aerodynamicists were anything but sure that the A-4 would not in that region become unstable and tumble. What would be the effect upon the rocket of the dreaded compression shocks? So little was known about transonic and supersonic flow at that time that we feared the A-4's thin skin might be stripped off like paper. But the A-4 confounded the gloomy pessimists and lived up to all expectations. Despite a few months of "bugs" and headaches, it was in mass production a year later.

The A-4 doesn't seem very huge alongside a modern global bombing aircraft, and I see no reason in the world why we cannot build vastly larger rockets. Of course, they will have the customary teething troubles, but they will also eventually meet their specifications and attain their design performances.

It is the inevitable teething troubles which seem to fascinate the critics of a coordinated space program, and they're always emphasizing them. "Sure," they say, "there have been a very few successful high-altitude rocket launching headlines in the papers, but how many attempts which failed? You don't hear much, if anything about them ..." I'll subscribe to the comment one hundred percent, for that's exactly how it is... But let's go a little deeper into the whole subject. Why is it so difficult to build a reliable large-sized rocket, let alone a series of them? Is it, perhaps, because of a myriad components, any or all of which might fail? That's hardly a satisfying answer when you consider that any multi-motor aircraft consists of a much larger number of components, to which thousands upon thousands of passengers entrust their transatlantic lives year in and year out. The real answer is a bit more involved: no rocket designer has yet had the pleasure and privilege of listening to the complaints of the test pilot who flew it for the first time. We have to rely upon indirect reports on flight conditions telemetered back to Earth, and it's a pretty tedious job to find out what really went on. And furthermore, before a rocket can telemeter back to us a report covering every phase of its flight, it has got to be at least a halfway reliable rocket! That is to say that its designer and builder must know in advance—at least approximately—what stresses each component of the rocket will undergo. He must have in mind how vibration is going to affect a certain relay during the flight, say; and roughly what the effect of transporting and setting-up will be on a given hydraulic line. He doesn't dare carry out an experimental launching until he has amassed a large collection of that kind of data from Earthbound trials. Thereafter, he must institute an extended flight test program during which the telemetering equipment will report to him what components are still faulty. That is the only way in which malfunctions may be discovered and corrected; it simply cannot be done by pre-study alone.

In order to develop a truly reliable large rocket—large as we think of it today, or "large" as it may become in the future—there is no substitute for a considerable number of complete test rockets. Otherwise bugs cannot be eliminated and weaknesses cured. The international airlines do not buy aircraft until the latter are in quantity production and have proved their reliability in thousands of flights. Any rocket development program limited to some 10 or 15 complete rockets is, therefore, hopelessly handicapped when it comes to achieving reliability. One or two individual rockets may make the headlines with incredible performances, but the design as such would fall far below the standards imposed upon a transport aircraft. Thus, if we want large rockets to achieve comparable reliability with the limited means available, it will be best to concentrate on a small number of designs, but to produce these in sufficient quantity to cull out all the bugs and thus eliminate all sources of malfunction.

What applies to a complete rocket in this respect applies to each and every component which goes to make up the whole. Just as in aircraft, the propulsion unit is one of the components of a rocket most subject to malfunction. For decades the aircraft industry has recognized the wisdom and practicality of installing the same power plant or propulsion unit in various types of airframes. The identical engine design will often be found in single-engine, twin-engine and four-engine aircraft and the advantage thus accruing to the industry is obvious, for each company benefits by the collective experience of all the others using the same power plant. Rocketry has yet to benefit by this simple and successful principle, for every existing rocket design uses its own, individual patent propellant combination and its own, tailored-to-measure thrust unit. And so you cannot today fire a rocket without simultaneously testing the propulsion unit to be used with it! Just imagine how hard it would hit airplane designers if they knew that the first test flight of a new creation was invariably to be made with inadequately tested, new-fangled engines! Fortunately for those designers, it is quite customary in the aircraft industry to test new power plants, preferably in airframes whose airworthiness has withstood the test of time.

Now there are, indeed, certain practical limits to the size of rocket propulsion units. Among the limiting factors is the necessary increase in wall thickness as combustion chambers grow larger; this multiplies the cooling problems. Test stands for such large combustion chambers are also extremely high in cost. I hesitate, nevertheless, to set a definite limit for thrusts of propulsion units, since I am convinced that any limit so set would inevitably fall below that which will be obtained within the next few years. Yet it is quite safe to predict that propulsion units of moderate size will be used in clusters when large thrusts are desired—just as multiple engines have been applied to large aircraft.

It may be objected that this practice will tend to further decrease the reliability of the rocket, but it should not be forgotten that, during the production of the required larger number of thrust units, there will be an opportunity to eliminate sources of malfunctioning which would not be available during the development of larger units in smaller quantity. Technical history supports the thesis that use of greater numbers of well-tested, standardized parts tends toward reliability of the completed machine, as compared to use of fewer, scantily-tested, tailor-made parts.

A multi-stage rocket allows us to take another step ahead and to utilize the same, well-tested thrust unit in different-sized clusters in the various stages. This will permit us to devote more money, brains and experimentation to the development of the standard thrust unit, since it will be utilized in even greater numbers than otherwise.

Let me cite still another example of the application of this principle in modern technology—one which touches the lives of every one of us and which is well-nigh proverbially successful—the modern automatic dial-type telephone central in my metropolitan city. There is probably no electromechanical installation in the whole world which is so dependent upon the proper functioning of literally hundreds of thousands of parts as that, vastly complicated as it is. Does not its proverbial reliability stand in direct contradiction to the complication? Not so at all; the fact is that each element composing it is made in such large quantities that the producers of those elements have worked them out to the last ten-thousandth of an inch and have shaken them down to perfection. When *all* of the elements of a complicated installation are practically trouble-free, the installation in its entirety also becomes practically trouble-free.

Now, if we want to tackle space travel along the lines of this "building-block" principle, we've got to lay out a definite course before we do anything else. If we propose to develop a truly reliable thrust unit which can ultimately be used in large clusters in the booster stages of a multi-stage rocket, we must channel our present development around a very few—the fewer the better—design types. We cannot put such a standardization into effect before we have made up our minds what propellants and what performances we desire for the future. If we ever intend to build a satellite rocket—even to seriously building one—we must today begin to lay out on paper the individual steps which lead up to it. We must transpose the results of our first, idealized optimum computations into structural designs of the intermediate steps. We must work up standards for the essential components, primarily for those making up the propulsion units. In other words, we must plot our future course! We need a coordinated space program and we need it *now*!

Some people maintain that to set up such a program today isn't scientific at all. They claim it would be speculation, pure and simple. But here's a statement from one of the outstanding scientists of our times; he defines the most important task of science as that of formulating predictions and conclusions for the future out of observations and experiences of the past and present. Now isn't that exactly what we would be doing if we set up a working schedule for the space program? Isn't it as plain as a pikestaff that any real space program is permanently blocked unless we get away from the frittering type of made-to-measure individual projects—mostly with inadequate means—which have monopolized the field up to now? As far as I am personally concerned, any fond belief that we can conquer space without a coordinated space project isn't a speculation—it's an illusion.

Getting back to cases, the same principle I've just discussed with respect to propulsion units also applies to automatic control mechanisms in rockets. Such automatic control mechanisms have the reputation of being not only extremely complicated, but likewise very susceptible to malfunctions, and in many cases these reputations are deserved. We should not forget, however, that many present-day military rockets are subject to extreme demands in the matter of guidance. Just consider for a moment the radar-guided ground-to-air missile which is required to nullify the evasive action of an enemy aircraft which is not even within the range of optical vision. And the accurate guidance of extremely long-range ground-to-ground rockets is just as difficult, or perhaps more so. Any attempt at standardizing the components of such guidance mechanisms—with the objective of securing improved reliability—is often faced by a basic problem, namely that the physical principles upon which the mechanisms are based must still be tested to determine whether they are competent to fulfill the high degree of exactitude demanded of them.

As compared with the guidance problems posed by such military rockets, the guidance of an unmanned satellite rocket is mere child's play. If we are not overly meticulous about the orbit in which we wish to fly around the Earth, we can assemble the guidance system of such a rocket out of components commercially available today. But even if we do insist upon higher accuracy, we're still far distant from any such micro-standards of precision as are called for in a military rocket of long range.

How about my unsolved *scientific* questions concerning the satellite rocket? Are there any such which could delay its development? It has been claimed that our engineers have drained dry the reservoir of basic research to such an extent that we should abandon all hope of any rapid progress during the next few years. Let us see whether we possess enough fundamental knowledge to schedule a comprehensive space program development and whether we may dare to begin the engineering phase of the first practical step, let's say an unmanned satellite rocket!

We already have definitive research and experimental aerodynamic data to assure us that we can send a satellite rocket into a circum-Tellurian orbit without encountering any insoluble aerodynamic problems. The reason for this statement is that we possess a rich mass of actual aerodynamic data up to about Mach 7 or 8, and what we know is that at such velocities an orbital rocket will have reached altitudes where the air is so thin that we may neglect aerodynamic drag and stability. The same data permits us to state that during the ascent of an orbital rocket there will be no serious skin temperature problems. None of the biological effects of space travel—namely extended weightlessness, ability of the animal organism to function under high Gs, or cosmic and solar radiation—have any bearing upon an unmanned satellite rocket.

There are, hence, no scientific fields in which there still exist unanswered questions which might nullify the prospects of success of an *unmanned* satellite rocket, but the story is a different one when one considers returning the top stage of a *manned vessel* safely to Earth. The primary unsolved question which must be answered prior to the design and construction of such a ship is the aerodynamic one, and that of heat transfer to its skin during the return flight. Parts of this flight will take place in the hypersonic region, i.e. at more than Mach 10. Here we are faced by a lack of some of the data affecting the design and construction of the winged airframe constituting the top stage. It must be controllable in the hypersonic region and be capable not only of withstanding extremely high skin temperatures, but of protecting the human life within. We do have certain concepts as to how to successfully tackle these difficult problems, but the developmental engineers urgently need a good many more data than those with which we can today furnish them. Furthermore, a manned satellite rocket faces the *biological* problems hereinbefore touched upon, not to mention a multitude of others related to the effects of impact of micrometeors, meteor bumpers and the like. There is also the extremely important and involved question of *training* the crew and providing for them satisfactory life-saving and emergency equipment.

We definitely need accurate information in order effectively to attack the above-mentioned phases of space travel, but we surely cannot obtain it by sitting back and bewailing the fact that the time is not yet ripe to do more than think about them.

The first step towards solving the problems of hypersonic flight may best be taken by utilizing existing supersonic aircraft in combination with jettisonable acceleration stages. Such a technique would enable us to make manned experimental flights in the region up to Mach 10 within a relatively short time. Control and aerodynamic problems relating to such flights may be examined in existing wind tunnels, and the full-scale effects of aerodynamic heating may be approached gradually by flying existing aircraft at the higher Mach numbers—first at extreme altitudes, where there is less heat transfer, then descending to regions where the heating effect is greater. The reliability of the equipment, in particular that of the jettisonable thrust stages, would be obtained by utilizing such standardized thrust units as I have earlier strongly advocated. You will appreciate, I am sure, that this part of the development program, namely manned hypersonic flight, is only possible if it is made part and parcel of a coordinated space program.

Of course training and life-saving equipment for space crews must be developed and tested, and this may easily be combined with this phase of the development program. Pressure suit and ejection parachute seats are already well recognized and tried safety devices in the field of high-speed aircraft testing and operation. It is no far cry from these devices to the ejectable pressurized capsule which will bear a crew member of a future manned satellite rocket to safety. Nor is it an overwhelming problem to locate a device essentially similar to present-day flight simulators within a centrifuge which will reproduce accelerations such as the crew will suffer during ascent to the orbit. To do this is but a logical extrapolation of present-day flight crew training methods which have been thoroughly developed and tested.

Undiscussed so far is the subject of extended weightlessness, cosmic and solar radiation, and meteoric dust in space. A modicum of knowledge on these important subjects can best be obtained by the use of an *unmanned satellite rocket*. Research has given us already all the answers we need to design, build and operate one. It has also been shown that we can well achieve a satisfactory degree of reliability in a satellite rocket if only we return to the badly neglected "building-block" or standardization principle.

Imagine, if you will, a three-stage liquid rocket some 46 meters high and 9 meters in diameter—that is to say, about 3 times as high and 5 times as thick as the old, familiar A-4. This rocket is expendable, and will be fired into a satellite orbit of approximately circular nature located about 320 kilometers above the Earth's surface. Whether or not it maintains its orbit with mathematical exactitude is not of the slightest importance. There will be no attempt at salvaging either of the booster stages. The third stage will contain a small 10 watt power plant, probably operated by solar heat. There will be one of the telemetering sets already in use in rocketry and aircraft testing and capable of transmitting to Earth from 60 to 80 readings of various instruments. There will also be a pressurized chamber containing animals such as monkeys. These will automatically be fed at regular intervals and their behavior under conditions of weightlessness will be televised so that observing biologists may take due cognizance of its effects. The third stage will also contain a battery of Geiger counters for registering the incidence of primary cosmic rays. These data will be telemetered to Earth over some of the available channels. Other Geiger counters, adjusted to greater sensitivity, will similarly record and transmit the Sun's x-rays. The skin of the third stage will be spotted with microphones capable not only of registering the impact of micrometeors, but also utilizing the familiar acoustic triangulation process for locating the point of impact.

We may install all the above-mentioned devises in a single satellite rocket, or it may be preferable to distribute them over a number of such rockets. This type of expendable satellite rocket constitutes actually an ideal means for examining those questions which cannot be approached in any terrestrial laboratory by the very nature of things.

It is a very simple matter to enlist the unmanned satellite rocket in our overall development and research plan. The first booster of the unmanned rocket may have a certain thrust—call it "B"—and we may compute the ensuing manned vessel as requiring identical thrust for its second booster stage. Analogously, we may design both manned and unmanned rockets so that the thrust "C" of the unmanned second stage equals that required for the third stage of the larger space vessel. The third stage of the unmanned satellite rocket may well have a propulsion unit of approximately the same thrust as that of the A-4. This gives us an opportunity to flight test our standardized thrust unit in the combinations "B," "C" and "D" when we are working with the unmanned rocket. When, at a later date, and in possession of the test data obtained from unmanned satellites and the manned hypersonic flight test program, we proceed with our manned satellite rocket, we may employ in it the already tested thrust unit clusters "B" and "C" for the two upper stages. Thus the only remaining development problem, as far as the power plants are concerned, is the combination of a sufficient number of standardized thrust units so as to form the cluster "A" for the first booster stage. The manned satellite rocket is thus integrated into the larger project, thus reducing its development cost, or looking at it from the other side of the fence, we develop the manned rocket by incorporating in it many of the elements which have already been flight tested in the crewless device. Either way you look at it, money and effort have been saved, while the reliability of the manned rocket has in advance been enormously enhanced.

Let me once more pose my original question: how far in the future is manned space flight? Unless we adopt a well-ordered, carefully-thought-out program, I say it is very far in the future— much too far! If we piously fold our hands in our laps while we await some apocryphal revelation of research, we may have to wait a hundred years before the first men circle the Earth in a satellite orbit! On the other hand, if we chart a careful course and stick to it, it will require but a few years for us to create the building blocks with which to construct the pyramid from whose apex man may leap into space! And we shall require the minimum of practical development work to do it. I am more and more convinced that in ten to fifteen years we can have not only a manned satellite rocket, but perhaps even a manned outer station: that is, if we stick to a well-charted program with the same perseverance and fixity of purpose which during the late war created the atom bomb and Peenemünde's A-4.

I would hardly dare to predict whether such a program can be instituted and carried through for year upon year during these piping times of peace, if such they can be called. That is a wholly different question which bears no relationship to technical or scientific matters. It is actually wholly dependent upon the urgency felt by the Western Powers regarding the necessity of the development, and upon the amount of money and energy those Powers devote to it.

I only hope that we shall not wait to adopt the program until after our astronomers have reported a new and unsuspected aster [star] moving across their fields of vision with menacing speed.

*** END ***

— Chapter 5 —

Cold War Paranoia & Sputnik 1

Never before had so small and so harmless an object created such consternation.
— Daniel J. Boorstein: *The Americans: The Democratic Experience*[32]

Between 1945 and 1989 the United States and the Soviet Union were in an intense political, military, and economic confrontation that came to be known as the Cold War. The struggle between the two superpowers dominated international affairs, and the conflicts it spawned raged across the globe. The world was seemingly divided into two armed camps: the United States and its allies against the Soviet Union and the Communist bloc.

The competition between the two superpowers was played out at many levels, but none was more visible, more consistent, or had greater impact on the United States than the arms race. It was a race driven by fear and fueled by uncertainty; a contest depicted by both sides as a struggle for national survival.

In retrospect, it is difficult to recapture the sense of fear and anxiety that, for many Americans, characterized the early years of the Cold War. From the United States' perspective, the Soviet Union and its communist allies appeared to be on the offensive around the globe. These were the days of the "Red Menace," a time when school children crouched under their desks during air raid drills, worried homeowners built fallout shelters, and the government conducted an intrusive campaign to ferret out shadowy "communist sympathizers" suspected of plotting against the nation.[33]

One of the many shock waves that affected the consciousness of the United States following World War II occurred August 1949 when the Soviet Union successfully exploded a nuclear device, thereby ending the American atomic monopoly. Many Americans jumped to the conclusion that the United States was losing the Cold War.[34] The Big Three alliance of World War II, often strained during the war itself, quickly splintered as the victors began to plan peace. Questions of territorial boundaries, spheres of influence, atomic weaponry, trade, economic reconstruction, political principles, and international organizations divided Britain, the United States, and the Soviet Union. Americans as a rule were resentful of the Marxist/Leninist teachings of Communism. Russia resented the fact that the United States refused to recognize the Soviet regime. Why the Cold War developed with such divisiveness is a topic of spirited debate among scholars. Recent scholarship makes clear that there is no single explanation for the origins of the Cold War.[35] Did Stalin and the Soviets represent a deceitful, dangerous expansionist threat? Or were they mainly motivated by understandable security concerns which prompted them to seek control over a ring of neighbors that had historically been a corridor for invasions? Were their actions manifestations of aggression, or were they reactions to a Western bloc attempting to impose its own influence, while flaunting the bomb on its hip?[36] There was/is a wide-ranging viewpoint that military power goes hand-in-hand with political power.

Aside from the various courses of action being presented by policy makers, probably many consider that the most compelling event was the Soviet launching of *Sputnik I* later in 1957. Not only did the Soviets have nuclear capability, they had the means to launch a possible nuclear warhead great distances.

"Listen now," said the NBC radio network announcer on the night of October 4, 1957, "for the sound that forevermore separates the old from the new." Next came the chirping in the key of A-flat from outer space that the Associated Press called the "deep beep-beep." Emanating from a simple transmitter aboard the Soviet Sputnik satellite, the chirp lasted three-tenths of a second, followed by a three-tenths-of-a second pause ... repeated over and over again.[37] In the history of space flight, there are few events that so affected the public; scientists, academia, engineers, and professionals in general could not believe that the simple, backward Soviets were capable of surpassing the United States in any field, let alone launching the first artificial "Moon" orbiting the Earth. In Huntsville, von Braun declared, "We could have done it with our Redstone two years ago!"

Air Force General James Gavin recalled to von Braun, prior to the launching of *Sputnik I*, "I had a hard time convincing Congress that launching a satellite was possible. Finally I decided to bring Dr. von Braun before a Senate Committee. I brought you into the hearing room where you began to talk about the Soviet capabilities. After listening awhile, Senator Ellender said that we must be out of our minds, that the Soviets couldn't possibly launch a missile or a satellite. He had just come from a visit to the Soviet Union and, after seeing the ancient automobiles—and very few of them—on the streets, he was convinced that we were entirely wrong. As you listened, you nodded your head, acknowledging what the Senator was saying and I was a bit concerned that the recorder of the hearings might record von Braun's gestures as agreement. So, I handed you a note suggesting that you be careful not to give the impression of agreeing with the Senator, since I knew that neither one of us did agree with him. The Chairman of the Committee brought the hearing to an end and then called me before him and threatened to throw me out of the hearing for attempting to influence a witness. It brought the hearings to an end and no one was convinced that the Soviets could possibly launch a satellite."[38]

On Friday evening, October 4, 1957, some fifty American scientists involved in the celebration of the International Geophysical Year (IGY) were attending a reception given by their Russian counterparts at the Soviet Embassy in Washington, D.C. A few minutes before seven, word arrived that Moscow radio was broadcasting the astonishing news that the Soviet Union had launched the world's first artificial satellite, which it dubbed *Sputnik*. The American scientists were caught completely by surprise. After a moment's consideration, one of the American scientists clapped his hands, asked for everyone's attention and raised his glass to toast the Soviets on their accomplishment, the launch of the first man-made satellite in human history—the true dawn of the space age. How could a backward Communist nation beat the United States into space?[39]

Sergei Korolev, Russia's enigmatic rocket engineer, waited ninety-three minutes before *Sputnik I* had completed its first orbit and ground control could confirm that it was indeed overhead, beeping. Then he phoned Khrushchev at the Kremlin. No one could have anticipated the media riot that followed *Sputnik* in the West. The panic was further exacerbated when details about the man who had orchestrated this magnificent celestial maneuver remained cloaked in secrecy. The West was only informed that the man behind *Sputnik* was known as the Chief Designer. While it jolted the rest of the world, the successful launch received casual treatment, in Moscow. Nikita Krushchev never dreamed it would have so powerful an effect. He called it "just another Korolev rocket launch."[40]

President Eisenhower's White House tried to minimize the significance of the *Sputnik* launch. He left for a weekend of golf at Gettysburg. Press Secretary, James Hagerty, and Secretary of State, John Foster Dulles, stated the *Sputnik* "was no surprise" and that the President was being kept informed.[41]

Michigan Governor, G. Mennon Williams, waxed poetic:

> *Oh little Sputnik flying high*
> *With made-in-Moscow beep,*
> *You tell the world it's a Commie sky*
> *and Uncle Sam's asleep*
> *You say on fairway and on rough*
> *The Kremlin knows it all,*
> *We hope our golfer knows enough*
> *To get us on the ball.*[42]

Senate Majority Leader, Lyndon Johnson, said the Eisenhower administration had made one of the most monumental political and foreign policy blunders in the history of the nation. *Sputnik*, Johnson reminded America, represented the high ground, mastery of the heavens. Maybe it was all right with others in government, he told reporters, but he for one didn't care to go to bed by the light of a Communist Moon.[43]

Senator Richard Russell, chairman of the Senate Armed Services Committee, said the Soviet satellite was "proof of growing Communist superiority in the all-important missile field. We now know beyond a doubt that the Russians have the ultimate weapon—a long-range missile capable of delivering atomic and hydrogen explosives across continents and oceans."[44]

James R. Killian, president of MIT and soon to become White House Science Advisor, wrote that *Sputnik* caused a "crisis of confidence" among the American people. The conservative *U.S. News & World Report* likened *Sputnik* to "the first splitting of the atom." Edward Teller, the "father of the H-bomb," told a television audience that the United States had lost "a battle more important and greater than Pearl Harbor."[45]

Fortunately for mankind, only the smaller missiles developed in the Cold War years have been fired in anger. The warheads of the hundreds of long-range missiles that stood poised in the United States and the Soviet Union could have killed hundreds of millions of people and destroyed a large percentage of the world's industry if they had ever been used. The tenuousness of the balance of power that prevented their use was shown in 1962, when the Soviets began setting up offensive ballistic missiles in Cuba. The world came close to nuclear holocaust in the confrontation between President John F. Kennedy and Premier Nikita S. Khrushchev in October 1962. There are still no guarantees that the holocaust will not be triggered by some confrontation in the future.[46]

Wernher von Braun was aware of the threat of living under a dictatorship, as was the plight he feared for the citizens of the Soviet Union. Von Braun knew the future for a world at peace was through dominating the space program. Unfortunately for him, the threat to the world by Stalin was not apparent until the Soviets launched *Sputnik*. Having lived under a ruthless dictator, the knowledge that the Soviets were working on missiles was of great concern to von Braun. He recognized in the Soviet regime the similarities with life under the Nazi dictator, Hitler.

In 1952, and the following thirteen years, his speeches reflected what he felt would be the fate of the United States and the rest of the world if the politicians and citizens did not become aware of the necessity for "Space Superiority" by the Free World.

Von Braun presented the following talk to the Business Advisory Council for the Department of Commerce in Washington, DC on September 17, 1952.

SPACE SUPERIORITY AS A MEANS FOR ACHIEVING WORLD PEACE [47]

THE MOST DISTINGUISHED MILITARY THINKING OF OUR TIME CONCEDES THAT THE only way to win a third World War is to prevent its outbreak.

The statesmen of our country have made every conceivable effort to ease the tensions arising from the dislocation of the balance of power which followed World War II, but their labors at the conference table continue to be bitterly disappointing. The net result of all the talk is a very expensive realization that there is only one way in which a treaty with the dictators of the East can be made to stick: namely to back it up with enough force to compel its observance.

Thus the West finds itself obliged to arm to the teeth in the interest of maintaining an uneasy peace in this tortured world and the United States bears the major brunt. In rearming ourselves we face a double problem. The first is to create a *deterrent power* which shall be sufficiently effective to inhibit the East from continuing its aggressive expansion ending in all-out war. Secondly, we must build up *fighting power* so that we may have the best prospects for success—and minimum destruction in our own and allied countries—if global war cannot be avoided. In the light of the introductory statement—credited to General Marshall—that to prevent such a war is to win it, it is only logical for the United States to give first priority and immediate, maximum attention to creating and making effective *deterrent power*.

At the present time our deterrent power depends upon the combination of atom bombs and strategic bombers of long ago. There is, however, considerable uncertainty as to just how much longer that combination will remain effective. The mere possession of atom bombs by the Red rulers will not render Red nerve centers and nerves less vulnerable to our bombs, but the question is whether our global bombers will be able to reach their targets when the time comes, if come it does.

But a short two years ago it was almost the consensus of opinion of engineer and military alike that certain large intercontinental bombers were immune to any and all attacks. Since that time, the Russians have not only developed, but put into mass production interceptors with a much higher ceiling and a top speed much greater than those bombers. Of course we have transonic bombers in the works, but nobody questions today that by the time we get them, the Russians will be all ready with supersonic interceptors. We also know that the Soviets are vying with us and other Western Nations in the development of new and effective ground-to-air guided missiles which could make the life in a heavy bomber crew anything but a bed of roses.

The atom bomber may have been the "ultimate weapon" heretofore, but this will not be the case much longer. Like the battleship, the atom bomber will become just another weapon, capable indeed of playing an effective role in war, but its deterrent power is on the decline. The handwriting is on the wall.

Winston Churchill thinks that the uneasy peace the world has enjoyed since 1945 has been due to the deterrent power exercised by strategic bombers with their capacity for delivering atom bombs anywhere they might be needed. It was this country's statesmen, her industrialists, her engineers, her designers, her scientists, her workmen, her airmen who had the vision, the industry and the initiative to bring that deterrent to a third World War into being. I might include the taxpayer, without whose contribution the enormous financial requirements of the Strategic Bomber concept and the Atomic Program could not have been met.

It is now the time, however, when we must familiarize ourselves with the thought that strategic bombing will soon be relegated to a secondary position, to the status of the battleship, so to speak. We must seek a new "ultimate weapon" which will preferably not only return to us that deciding "edge" we once had over Red aggression, but likewise be kinder to the taxpayer and be able to contribute something constructive to the world whose uneasy armistice we hope it will successfully transmute into permanent peace.

Rocketry is, I believe, capable of solving the world's peace problems more effectively than any other branch of science and engineering, and simultaneously—that is to say without additional expenditure—doing a great deal of advancement for mankind. The first nation to launch a rocket ship that is able to carry a crew out beyond the stratosphere, will posses what may well be the long-sought "ultimate weapon." But beyond that, and once it has fully exerted its deterrent effect upon would-be aggressors, it will be capable of serving an infinitude of scientific—that is to say, humanitarian—ends.

Ladies and Gentlemen: I want to appeal to you for your support of an orderly coordinated space program in this country. Such a program will require some few millions of dollars for studies to be carried on during its first phases, but the investment will be repaid a hundred times over when we reach the "hardware stage," and conditions will force us to accelerate that stage, mark my words! We've got mighty little time to lose, for we know that the Soviets are thinking along the same lines. If we do not wish them to wrest the control of space from us, it's time, and high time we acted!

*** END ***

– Chapter 6 –

Von Braun spoke at the Hayden Planetarium lecture series, on the early steps in the realization of a permanent space station on October 13, 1952. In one of the shortest speeches he ever presented, 473 words, von Braun put forward an outline for a trip to the Moon. NASA Headquarters senior scientist, Milton Rosen, was to present a paper as well: *A Down-to-Earth View of Space Flight.* This was a complete contradiction to von Braun's theories. Rosen, reminiscing of the event in 1952, told von Braun that, "Willy Ley, who was arranging the program, wanted me to modify or withdraw my remarks in fear that they might do damage to the cause of space flight. The Hayden Planetarium took the position I could say what I wanted, hence, a confrontation appeared to be imminent. But you were not concerned; indeed, you pointed out that if all of us sang the praises of space flight, the press would take little note of it. Your prediction was proven correct. 'Experts differ on future of space flight' was the headline on front pages of New York newspapers. Moreover, the conflict was featured in *Time* magazine."[48]

ACROSS THE SPACE FRONTIER [49]

WITHIN THE NEXT 25 YEARS, MAN WILL BE ABLE TO SET AFOOT A FULLY-FLEDGED expedition to the Moon. The rocket ships of this expedition will not take off from the Earth for a direct flight. They will be assembled in a satellite orbit leading around the Earth at an altitude of 1075 miles, to which the prefabricated parts of the ships will be carried in large, three-stage "satellite rocket ships." Unlike these satellite ships, which will be beautifully streamlined, the Moon ships will have an odd and ugly configuration, since streamlining is not necessary in airless outer space. The first lunar trip will *not* include a landing on the Moon. It will rather be a flight *around* the Moon. During this first trip the unknown back side of the Moon will be photographed and a suitable landing site for the lunar expedition determined. This landing site will probably be located not too distant from the Moon's North Pole, in order to avoid the high daytime temperatures prevailing near the Moon's equator.

The first actual descent to the Moon's surface will probably be made with three rocket-ships, two of which will be built for the round trip between the satellite orbit of departure and the Moon, whereas the third ship, designed for a one-day trip only, will be abandoned on the Moon's surface. The one-way ship will be capable of carrying approximately 270 tons of cargo to the Moon, which will enable the expedition to set up camp facilities for a stay of six weeks. Such cargo will include tracked vehicles for a thorough ground expedition of several hundred square miles of the Moon's surface.

The pressurized huts of the expedition headquarters, assembled from prefabricated parts, will probably be erected in a crevasse for maximum protection against meteorites and cosmic radiation. The tasks of the expedition will include geological studies and it will be attempted to determine the origin of the Moon. With the help of artificial "tremblers" produced by impact explosions of ballistic ground-to ground rockets fired from one point of the Moon to another, and by seismographic measurements of the travel of the shock waves through the interior of the Moon, the explorers will attempt to find the answer to the old question: whether, in prehistoric days, the Earth itself gave birth to the Moon, whether the Moon was a traveler from outer space which was "captured" by the Earth, or whether, as some more recent theories have it, the Moon actually constitutes a mass fused from cosmic debris.

All expedition members will return to the orbit of departure in the two round-trip ships, whence they will return to the Earth's surface in the winged upper stages of the three-stage satellite rocket ships.

The Moonbound and round trips will take five days each, whereas the expedition will stay on the Moon for six weeks.

*** END ***

– Chapter 7 –

The Associated General Contractors of America
Miami, Florida, January 1959

Mr. Chairman, distinguished guests, members of the Associated General Contractors:[50]

The relaxed and pleasure-filled setting of this meeting is quite different from that which surrounds my usual business visits to Florida. As you have probably gathered from the headlines, we fire our bigger missiles at Cape Canaveral, a few hundred miles to the north. The atmosphere there on such occasions is usually charged with tension and anticipation, and the missiles are more abundant today than the famed Cape mosquitoes!

I can remember the early days at Canaveral when our Army Redstone missile was the only rocket tested there, and when the entire operational strength of the Air Force and the Army engaged in this program squeezed uncomfortably into a single hangar. In those pioneer days we resorted to all kinds of expedients because money was so tight. I recall one occasion when we rushed into Cocoa [Beach] and bought sewing tape, thread, and needles in order to get a missile off the pad.

Today the Cape is dotted with enormous hangars, with concrete blockhouses of all sizes and shapes, with giant service towers and elaborate tracking stations. It is liberally supplied with the most advanced types of electronic and mechanical gear, which is utilized by all three of the Armed Services in the flight testing of big rockets.

I am sure the membership of your Association has contributed to the rapid build-up of facilities at the Cape, and also at our headquarters in northern Alabama, at numerous West Coast installations, and at missile contractor plants throughout the country. When we undertook to expand the development and testing facilities at Redstone Arsenal, the construction projects in some cases involved requirements novel to your industry because there had never been such buildings or test stands in this country. Much of what goes into rocketry is new; for much of it there is no experience. But it means the translation into concrete and steel, other metals and chemicals of new knowledge, and of satisfying requirements beyond the performance capabilities and strengths of conventional materials and structures. I might say that our experience to date with your industry has been highly satisfactory; your members have met the challenge.

I would like to be in position to say the same thing about our missile and space programs across the board. But I make it a practice to stick to the facts, however unpleasant they may be. At this date, in the second year of the Space Age, we must face the situation and admit that we have not yet met the challenge of a rival technology. We are moving in that direction, of course, but unfortunately we still cannot claim that we have attained parity, much less superiority.

Six months ago I attended the meeting of the International Astronautical Federation in Amsterdam, where scientists from the Free World and from countries behind the Iron Curtain exchanged information and gossip. These proceedings related to the International Geophysical Year program, by which we sought to gather more and better information about the planet on which we live, [and] upon its environment outside the sensible atmosphere. I say "newer" advisedly because, while outer space has been there for a long, long time, it is only within very recent years that we have begun to acquire useful and dependable data about its nature as a future stomping ground for men. There are still gaps in our knowledge, but thanks to IGY projects, such as scientific satellites and deep space probes, we have more information today than has ever been available to science.

I mention the Amsterdam meetings for two reasons. One, to point out that it demonstrated the advantage of mutual exchange of information between scientists regardless of their political affiliation or lack of it. Second, because from representatives of the Soviet Union who attended we learned that the next major project in their space program would be a lunar or space probe. Consequently, the event of early January did not come as a surprise. Having previously demonstrated the rocket propulsion power available by the *Sputnik* launchings, the Russians obviously possessed the capability to carry out the kind of project which has placed an instrumented, manmade planet in orbit around the Sun.

At the same international talkfest, and in various publications reported in this country since then, the Soviets plainly indicated that they are proceeding at full speed in a massive effort designed to retain and increase their alleged leadership in rocket technology. This is no longer subject to speculation. It is simply a matter of trying to keep up with the daily news and the repeated pronouncements from the Kremlin about interplanetary space flights and manned expeditions to the Moon. Soviet propagandists have never hesitated to tell us all kinds of stories when it served their purposes, but we should recognize that in the technical fields Russian performance has generally lived up to the claims made for it. I can only repeat what I said in the wake of the *Sputniks*—that our people must be prepared for further demonstrations of Russian competence in due course.

It would seem obvious that we must avoid complacency, of taking for granted American supremacy in all fields of human endeavor, of assuming that whenever we wish to do so we can overtake the Russians in rocketry. This can be dangerous when you stop to consider the possible implications. At best it is foolhardy. At the same time I do not agree with alarmists who cry that "we have lost our position as a first rate power," because that kind of sweeping generalization can result in hysterical, ill-conceived reaction, which can also be harmful.

We enjoy superiority in many facets of our society over anything the Communist system has yet produced, but the trouble is that we are prone to assume this superiority exists in all areas. It does not.

I suspect that we have not yet fully awakened to the breadth and depth of the Soviet challenge. There seems to be an illusion that this is a kind of race between competing teams of rocket builders, and that the "horse" we selected has pulled up lame in the stretch. Actually the Soviet effort antedated that of the United States. As a consequence, the competition got far down the track and into the first turn before we left the gate.

The notion that we must catch the Russians in this race, but have nothing else to worry about, is completely fallacious. A determined and able rival has challenged us in many areas—political, scientific, ideological, cultural, economic and educational, as well as technological. The Communist objective in this tremendous effort is by no means new; it is world domination now, as it has been since this totalitarian ideology succeeded in grasping power. What is new is the capability which the Soviet State has achieved and which it is building up by a massive educational program designed to provide a reservoir of scientific and engineering talent.

Apparently the Kremlin has been persuaded by the presence of NATO to pull up short of all-out war in its strategy of conquest. This simply means that in all other than purely military areas the attack is being pushed relentlessly. The test therefore is not confined to the abilities of Tom Jones, American rocket engineer, and Ivan Podbroasky, Russian missile designer. This is a test of stamina. ingenuity, courage and faith between every man who ever operated a lathe, who lays brick, who pours steel, who mixes concrete, who sews cloth, who analyzes chemicals, who solders connections, who resolves mathematical equations, who writes editorials, who comments about news, who delivers babies, who teaches children, who grows crops, who manages industry, who engages in trade and commerce, and his Soviet counterpart. This is not a contest of diplomatic skill between a small group of men in Washington and a small group of dictators in the Kremlin. It is a fight in which all of us are directly involved and the stake is our and our children's future. The result will demonstrate to all the world whether we are in fact, as we still have every reason to believe, the greatest nation on Earth and the shining hope of men everywhere.

Having lived under a dictatorship equally as cruel, equally as ambitious as is the Communist scheme of things, I know the price of lost freedom; I don't want my children or yours to pay it because of our failure to recognize the seriousness of the threat as we enter an Age of Space which can be the most fruitful and the most enjoyable of all the periods of recorded history.

There can be no possibility of mistake in evaluating Soviet intentions so far as outer space is concerned. The objective is to master the spatial environment, certainly for peaceful, scientific purposes initially, but with the long-range motive of putting the capability of manned travel through outer space to use for the same overriding goal of expansion of their sphere of influence which is the ultimate purpose in all Communist endeavors.

Let me briefly discuss the long-range space program which has been announced by the President of the Soviet Academy of Sciences. These are the various, possibly sequential, steps which he described:

First – satellites of such lifetime as to be practically permanent orbiters.
Second – recoverable satellites.
Third – manned satellites.
Fourth – rocket flights to the Moon and other celestial bodies.
Fifth – satellites of very high apogee orbits.
Sixth – interplanetary space stations which could support a considerable number of personnel over extended periods of time.
Finally – manned flights to Venus and Mars.

It should not be concluded that all of these projects are scheduled tomorrow in the Soviet timetable. Some require further advances in rocket technology because the entry of man into space and his safe return to Earth calls for the solution of many perplexing problems. While none of these projects are patently hostile, many of them could serve immediate military purposes. For example: reconnaissance on a world-wide basis; the delivery from space of destructive devices which could be guided to targets on Earth; the interception of orbital vehicles launched by others; the control of weather and similarly potentially aggressive employments.

But even more important, in the long run at least, seems to be the plain fact that if we don't match the ambitious Communist program to visit the Moon and the planets with an equally determined United States space flight program, we may in the not-too-distant future be surrounded by several planets flying the hammer and sickle flag.

It must also be remembered that it is not the scientist whose investigations make these things possible, not the engineer who builds them, who can decide how fast all this can be accomplished. That decision depends primarily on the funds made available for these objectives, and in the case of the Soviet Union, it will be made by a few men who consider it in the best interest of their long-range Communist expansion plans to vigorously pursue such an endeavor.

In a free country we can muster the necessary resources for a comparable plan only if the people themselves want it, support it, and are ready to make the necessary sacrifice.

I realize that this may sound grim and foreboding, and not at all in tune with the friendly atmosphere of this meeting. But it seems to me that the times call for sober realization and for a resolution to spit on our hands, pull in our belts, and get to work! The question of how much an effort will be required is not one for me to answer; I can only repeat that this involves every component of our civilization—the church, the schools, industry, science, the professions, as well as government. Perhaps there is a need for coordination of effort and contributions outside the framework of Government, which would bring together leaders in all fields and thus to get the message across that this contest will demand our best from each of us as American citizens.

In the days immediately ahead, as in the wake of the first *Sputnik*, we are bound to experience another period of soul searching, of close examination of our Government-supported programs, and of critical scrutiny of any area which is suspected of weakness or of performance below that required by the situation. Trying as these matters may become, they are very much a part of our way of life. It is to be hoped, however, that we will not spend too much time and effort in searching for scapegoats and thus lose sight of the other horse who is rounding the straight-away.

I would not have you think that nothing has been done by this country since the Russians made their opening bid. A great deal has happened in the last 12 months. Two new agencies have been established to carry forward our space programs—the Advanced Research Projects Agency of the Department of Defense (ARPA), which is concerned with space programs of military significance, and the National Aeronautics and Space Administration (NASA) which is directing the national [space] effort in general. The objectives of the United States have been outlined by President Eisenhower in clear and unmistakable language. Discussing the opportunities which a developing space technology can provide to extend man's knowledge of the Earth, the solar system, and the universe, Mr. Eisenhower said, "These opportunities reinforce my conviction that we and other nations have a great responsibility to promote the peaceful use of space and to utilize the new knowledge obtainable from space science and technology for the benefit of all mankind."

During 1958, all three of the Armed Services contributed to the International Geophysical Year program, the most spectacular evidences of which were the satellites and space probes. The Navy launched one satellite which will remain in orbit for years. The Army launched three, one of which, after surrendering most valuable scientific data, reentered the Earth's atmosphere and burnt up, and two of which are still in orbit. The Air Force launched its huge "talking Atlas satellite" and one of its space probes succeeded in penetrating more than 75,000 miles into outer space, while the Army's later space probe traveled some 65,000 miles before it, too, reentered the atmosphere.

Out of these projects we have derived a wealth of data about the spatial environment. An outstanding discovery was that credited to Dr. James Van Allen of the State University of Iowa, whose instrumentation in the three Explorer satellites and the *Pioneer III* space probe measured and defined the two doughnut-shaped bands of intense radiation activity which exist in space and which pose a definite hazard to manned space vehicles. We believe it will be possible either to steer a safe course around these bands, now that Dr. Van Allen has described their limits, or to provide the necessary shielding to protect a human being in a space vehicle from dangerous exposure to this kind of radiation. We have demonstrated techniques, relatively simple in fact, by which to control temperatures in orbital bodies to an acceptable range. We have pretty well satisfied ourselves that micrometeorite showers do not pose an insuperable risk. We have demonstrated highly-efficient tracking and communications systems, and more powerful systems of greatly extended range capabilities are now available.

In essence, each of the payloads hurled into space by U.S. rocket systems in 1958 were small-sized scientific laboratories equipped with highly efficient sensing and measuring devices and with transmitters which collected and reported back to Earth receiving stations data not otherwise obtainable about the spatial environment. More of these experiments are scheduled under NASA auspices in the coming year. Other space projects are planned which will not be advertised in advance of their occurrence.

I want to mention, in passing, one of the less obvious but highly interesting aspects of our achievements to date. Just one of the Earth's satellites, the *Explorer I*, has traveled approximately 130,000,000 miles since it was injected into orbit on January 31st from Cape Canaveral. If you consider the cost of the device in terms of distance it has traveled, and the millions of miles of additional distance it will cover before it eventually dies, I think you may concede that such a satellite is more economical transportation than your family car. My present guess is that *Explorer* will do better than 500 miles to a gallon, or the cost per mile will be reduced to something like one-tenth of a cent.

There are many intriguing aspects to this space business. For example, a small fleet of satellites equipped with modern electronic recording equipment could easily handle the entire mail volume of the world— [a] Postal Service, acquiring radio messages from Miami, for example, and relaying them to correct destinations in New York, San Francisco, London or Bombay. Worldwide television and radio relay can be achieved with similar satellites. They would be positioned at the correct altitude above Earth so that they remained in a relatively stationary position. Weather observation and forecasting can become an exact science with the utilization of orbiters, and the savings in lives and property damage by advance hurricane warning would in all likelihood far exceed the cost of the service.

Beyond these illustrations, the possibilities are more limitless than the human imagination. No one accustomed to bound concepts can possibly visualize all the wonders which may be the fruits of exploration into the far reaches of the universe. But I believe that enough can be visualized to justify the assertion that we are on the verge of a truly golden age of discovery, one that can yield manifold benefits if we can proceed in peace and without the menacing possibility of employing space technology for aggressive ends. The only sure guarantee of eliminating that threat is to advance our space exploration programs with redoubled vigor to the end that we establish, beyond all doubt or question, the supremacy of free peoples in this most challenging of all fields.

*** END ***

— Chapter 8 —

Needless to say, von Braun was an eager passionate student when his studies involved astronomy. Fortunately for the three von Braun boys, their parents were able to send them to private schools that suited their interests. Mrs. von Braun recalled about Wernher, "Learning was always easy for him, when he was ten we decided to send him to the Französiches Gymnasium (French Gymnasium) in Berlin to give him a somewhat broader basis of education. Around that time I asked him once what he would like to do later in life. 'I want to help turn the wheel of progress,' was his answer."

"Youth Faces Tomorrow" [51]
Chicago School Principals
(Student Science Fair) 16 January 1959

I MUST CONFESS THAT AN INVITATION TO ADDRESS A GROUP OF EDUCATORS presents a neat dilemma. It is much easier to speak to scientists, engineers, businessmen or industrialists about rocketry and outer space, because these audiences accept my statements and comment without suspicion of my motives. Experience has taught me that quite a different kind of reception awaits the speaker who ventures opinions in the field of public education. If he applies the broad brush of approval he may please his audience at the cost of his integrity. If he attempts anything like a critical evaluation, he risks displeasure or worse. I find this puzzling, for it is entirely logical to expect a keen interest in education from an informed citizenry. I fail to see any reason why education, per se, should be exempt from the kind of evaluation accorded all other publicly supported institutions in our society. There are many reasons why schoolmen should welcome that interest, not the least of which is that the better informed people are about educational problems, the more whole-heartedly they will support intelligent and progressive programs.

In company with many thousands of parents, I am aware of the wave of harsh criticism which has engulfed education in the past year. It stemmed from the sudden realization that a totalitarian state had managed to steal a march on us in technology, a field of endeavor which we had jealously prized as our private preserve. The breast-beating which followed has undoubtedly caused some injustices, but it has also produced beneficial results, including the recognition by the Congress that federal support must be more liberally provided in educational areas if we are to pull abreast of our competition and eventually forge ahead.

A few indices may be helpful at this point. The Soviet Union has caught up to us numerically in the number of scientists actively engaged in State-approved programs, and is graduating twice as many engineers annually—70,000 of them in 1958 compared to 32,000 in this country. There are more doctors in Russia than in the United States, approximately 164 per 100,000 persons compared to 130 per 100,000 persons here.

On the surface that makes us look pretty bad. It is necessary to keep a sense of proportion when discussing these things—there are other areas of vital importance to human welfare in which the Soviet Union is still running a poor second. For example, how many ministers of the Gospel were produced by Soviet theological schools last year? How many manufacturers, distributors and advertisers solely devoted to the challenging task of raising the standard of living of the population at large does the Soviet Union have, compared with the United States? How many independently managed newspapers, radio or television stations were functioning in Russia in 1958? I am sure that it is generally realized that Russian progress, admittedly substantial, has been concentrated exclusively in areas selected by an all-powerful state for objectives related to the spread of Communism at whatever cost to the Russian people.

The quick and easy assumption—that public education in the United States is somehow to blame for all our ills—does not square with the facts, and does great injustice to the dedicated men and women who try their best to prepare youth for tomorrow's world. If the schools were as inept as some critics assert, how can they explain our preeminence in so many fields of human endeavor? How do they explain the existence of cities like Chicago, or our national transportation systems, our great industrial complexes, our high standard of living which moves upward each year?

Even a cursory examination of the record gives the lie to the blanket allegations of incompetence or worse. Let me mention a few of the world-famous men, expert in science and technology, who have come from the American school.

Wilbur Wright attended high school in Richmond, Indiana. His brother, Orville, was a student in high school in Dayton, Ohio. Neither received the usual diploma. Wilbur just didn't bother to go around on commencement day to pick his up. Orville chose a specialized curriculum in his last year, which cost him the usual certificate. Note that he wasn't expelled because of it, but the statutory requirements had not been satisfied. Perhaps this indicates a need for reexamination of the criteria applied to preparation for graduation. At any rate Wilbur and Orville Wright designed and built the world's first successful power-driven, heavier-than-air flying machine.

Charles F. Kettering, who developed the first automobile self-starter and added many other inventions of great usefulness, was a product of public schools near Loudonville, Ohio, and of Ohio State University. He overcame a troublesome eye condition to complete his formal education.

Another Midwest youth, Harold C. Urey, attended public schools in Walkerton, Indiana and later studied at the University of Montana, the University of California and Copenhagen. He won the Nobel Prize for his discovery of heavy hydrogen (or deuterium) and is internationally recognized as one of the outstanding scientists of the 20th century.

Lee DeForest, born in Iowa, studied at the Sheffield School of Yale and patented 300 inventions, the best known of which is the vacuum tube. He was a pioneer in wireless telegraphy and radio broadcasting, which have contributed so much to breaking down the barriers between peoples of all races.

A contemporary figure is that of Dr. Jonas E. Salk, developer of the world-renowned anti-poliomyelitis vaccine. He attended New York City schools, helped pay for his education by working after school and by winning scholarships. He was graduated from the City College of New York and the New York University of Medicine.

These are but a few of a host of outstanding Americans of this and earlier generations who have made significant contributions to science and technology. Many others could be named who enriched our lives in other kinds of activity—industry, the arts, business and government. Certainly the American system of public education has no apology to make if we measure results in terms of the great figures.

But that recognition does not answer the question bothering many of us who are deeply concerned about the nation's future at this point in history. The rapid progress of science and technology demands not only that we have more Salks and Wrights and Ketterings, but that more intensive training be provided to more numbers of students than ever before. We must recognize individual abilities at the earliest possible ages and then cultivate and encourage them through the years of preparatory education to make it possible for them to achieve their greatest potential in adult life.

Should we devote more effort and attention to the slow child who needs it, or to the bright child who can put that extra attention to better use? The heart dictates that we spend more time with the slow child—every mother does it. But it is a question of national survival that we provide better opportunities for the exceptional child. For it is the gifted child of today that will be tomorrow's leader in our fateful struggle for scientific and technological superiority.

I believe the schools of Chicago reflect the awareness of your people and your school administration of the need. It is indeed gratifying to learn that the study of science and mathematics is mandatory for graduation here. You are nearing the end of the first year of this policy; it should continue through the years to come because it will immeasurably benefit your children and your city. It was a revelation to me to be told that elementary science and mathematics are introduced at the kindergarten level in a system which has 345,000 pupils in elementary schools and 106,000 in high schools. It is my opinion that this is the way by which to encourage a basic understanding of natural laws, which is essential to the citizen of today, and may become vital to the survival of the citizen of tomorrow.

I could wish that every city followed the example of Chicago in encouraging high school students to study science and mathematics throughout their four-year courses. What is even more praiseworthy is your policy of making available five years of study in these fields for advanced students who have shown they can handle more difficult work. Obviously Chicago was one of the first of the major cities to recognize the keener interest of youth in technical preparation, and it is to your everlasting credit that you did something about it.

We who are on the outside of the business of education are prone to fall into the trap of generalization. The fact is, of course, that American schools are quite individual and that, in spite of the praise or abuse heaped upon the John Dewey tradition, there is no single philosophy representative of all our educational systems across the land. I suppose it can be said that we intend to provide equal opportunities in education to all children, and further that we accept the responsibility of providing the necessary facilities and faculties by taxation. But if you weigh that statement against the multiplicity of solutions practiced throughout the United States, you will find all kinds of variances. It is almost impossible to select two public school systems at random and say, "These are identical and therefore representative of what we stand for." There are differences dictated by local circumstance and tradition; there are variations expressive of the individual philosophies of superintendents, principals or teachers. If there is any one factor characteristic of all our schools, it is this quality of difference.

I suspect that many parents and other well-intentioned citizens have come to appreciate that fact of late. They have learned to their chagrin that it is not quite as easy as they expected to introduce more of this, or to remove some of that, in the hope of coming up with a better solution. They have discovered a legal barrier in the form of state laws or local regulations, and they have found that others have preceded them and written their personal interests into the statutes—some of which still reflect concerns which were logical in an agricultural economy but have little validity in an industrialized community.

In some cities it is possible to earn high school diplomas without ever being exposed to mathematics beyond elementary algebra or geometry, without studying any of the physical sciences, without even a year of a foreign language. Not only is the student able to complete secondary schooling and win a diploma under such conditions, but in many areas he could not elect to study higher mathematics and the sciences if he wished to do so—the courses just aren't available to him. Apparently all that is expected of the student is that he will keep out of trouble for four years and compile a minimum number of points or credits. He must take a few required courses, perhaps a little English, a little U.S. History, or something called hygiene. For all the other hours spent in classrooms he has pretty free choice of what to do with his time. Left to his own devices, he naturally gravitates towards the easier subject—why bother with the difficult? When this circumstance exists, it does little good to point out that the student in Russia spends from 32 to 36 hours in class against the 20 to 25 hours required in U.S. secondary schools.

I doubt if that kind of easy choice was open to the high school pupils of the era which produced the Wright Brothers! Common sense argues against permitting a boy in his early teens to select the kind of preparation he needs for his future, when he does not even know what that future will or could be. I can assure you that I needed some pretty strong guidance at that age. Why not carry this present laissez faire approach to its logical conclusion and allow the baby to eat whatever it will, or permit a youth to play ball rather than waste the daylight hours in school?

We hear much these days about the importance of challenging youth, of awakening them to the immense possibilities opening before us as we probe a new dimension, that of outer space. My personal mail, running into the thousands of letters a month, from boys and girls in all sections of the country, convinces me that youth recognizes the challenge. I am not so sure, however, that we oldsters know what must be done to meet it. Look at the temptations to which a young fellow is subjected in our booming economy. He knows that he can sell cars, or run a filling station, and make more money than his teacher ever will, without benefit of college education. Chances are that if he makes the choice, and has the necessary stamina and willingness to work, he'll be living in a better home and playing golf at a more expensive club than his classmate can afford who went on to earn an engineering degree and is now developing guided missiles or TV sets. Here is an aspect of the problem that is clearly not the responsibility of the public schools. Rather, it is something which concerns all of us who want to see this nation grow and prosper, and it comes down to the brass tacks issue of what kind of rewards we offer to "eggheads" and intellectuals and scientists and engineers, as well as school teachers.

Make no mistake about it; the future of America in an age dominated by technology will depend largely upon the quantity and quality of engineers and scientists available to carry the nation forward. Thus we must face up to the situation and ensure that adequate recognition and reward are provided to them and to the teaching profession which prepares them. Perhaps then the embryonic Salks and DeForests will get more encouragement from dads who come to realize that there is something more to college than football and basketball, and who understand that graduate engineers and physicists will command remuneration comparable to that of the M.D. or a stock broker.

I believe, too, that we need to ponder carefully the perspective with which we look upon these concerns. Money has become too much the measure of success and achievement to the people of my generation, and that attitude echoes in the things which interest our children. I doubt if the great discoveries have derived from efforts directed solely towards the profit side of the ledger—many of them resulted from the labors of researchers and inventors interested only in knowledge and not the ups and downs of the stock market. There is virtue and reward in achievement for the sake of achievement, which may be an old saw, but one which needs resharpening and more use.

It is not financial reward, certainly, which motivates the rocket developer; yet he is constantly asked, "why do we do these things? Will they ever pay off?" Can you imagine anyone asking that of an Einstein or a Pasteur? There will be payoffs, of course, as we advance into space, but the most important, the reward of greatest lasting value, is our increased understanding of the spatial environment and of its influence upon our Earthly existence. You cannot put a price upon such a discovery as that of the Van Allen Radiation Belt, made possible through the launching of several Army satellites of relatively small weight. We have opened the door into the limitless reaches of the Universe by such modest beginnings and we can see just far enough ahead to know that man is at the threshold of a momentous era. Here is opportunity, challenge, adventure so tremendous as to beggar description. Here is the tomorrow which youth wants to embrace and for which we must provide whatever guidance and preparation we can. I am confident that as more of our public school systems initiate programs like that which you are carrying forward, we shall begin to prepare for the new age.

*** END ***

– Chapter 9 –

Women's Forum on National Security
Washington, D.C. 30 January 1959

The Women's Forum on National Security was comprised of over two million women through membership in fifteen participating organizations. Membership eligibility was rooted in the military services of relatives and descendants. The purpose of the Forum was to inform, arouse and activate public opinion among the women of the country as it related to the varied aspects of National Security. The theme of the conference was "Changing Dimensions in National Security." In 1959, that was a timely subject, not only for security issues, but for women joining the work force. The main security issue was the "Communist Threat." As von Braun writes:

Women's Role in a Changing World [52]

WE HAVE ABRUPTLY ENTERED UPON AN ERA IN WHICH MEN MUST TRY TO RELEASE their imaginations to the contemplation of limitless distances and worlds beyond our own, to new phenomena which have no counterpart on Earth, all of which have suddenly been brought much closer by the medium of rocketry. It is entirely logical, therefore, that you should be seeking answers to questions inspired by these events. I suggest you consider two possibilities in national security:

> First – accepting the definition of security as "freedom from danger," we must seek to achieve that objective by eliminating the danger.
>
> Second – if we find that we cannot eliminate the threat, then we must take every precaution to insure that our protection is adequate in all circumstances.

The problem of adequate security is complicated by the fact that it cannot, in an era dominated by technology, be a passive state. It must be treated as a constantly changing situation, directly influenced by technical progress. Suppose that your daily newspaper reported a satellite was orbiting the Earth and crossing this continent with monotonous regularity. Its trajectory carried it over major centers of population, great industrial complexes, seaports and military bases. This sort of thing has occurred within the year, so it does not require any stretch of imagination to expect it to be repeated. Suppose that the satellite carries electronic devices and transmission equipment by which it reports visually to the base from which it was launched everything that is "seen." How much secrecy remains to us in the conduct of large-scale defensive operations?

Rockets capable of injecting such satellites are available behind the Iron Curtain, as well as in this country. The mechanics of disarmament are much more complex than was the case in the 1920s or in the years following World War II. A nation covetous of others might subscribe to the reduction of conventional armaments and make a good show of it. Thus the danger of war in recognizable guise might seem to be reduced, but the same power could, under the mask of scientific research and development, retain the potential for destruction and conquest by the use of space vehicles for these ends. The long-range ballistic missile, in modified form, becomes the carrier by which to propel objects into space. We cannot achieve a state of complete freedom from danger, however desirable it may be, until men everywhere have convincingly demonstrated their good will and their intention to live together in amity, renouncing aggressive designs upon their neighbors. Until that day dawns, the second alternative is the only course open to us—to protect our country against *any conceivable threat*.

In the years which have elapsed since the last great war, 700,000,000 Asians have fallen into the Communist orbit and 100,000,000 people in Central Europe have been swallowed up while the Soviet Union has forged ahead to expand its industrial, economic, military and political strength. Today a tightly-knit Communist empire stretches from Berlin to the South China Seas, challenging the Free World at every turn.

A French writer, Amaury de Riancourt, addressed himself to the question of what has happened to the West that this disastrous change in the balance of power, which has in part halted mankind's progress towards the better life, should have come to pass with counteraction. He commented that the "answer stares at us; it is so obvious that we simply overlook it. The answer is that the Communist world has a strong faith to live by, an overall philosophy to guide its actions, a definite goal to look forward to and a will to achieve it." By contrast, he believes the West has turned its back on its past and has no spiritual aim.

I do not quarrel with his explanation of Communist gains—the facts support it. I believe that in this summary of the situation we may find a challenge to the women of America who seek their proper place in a swiftly changing world. It takes a certain amount of courage, of course, or just plain foolhardiness, to tell any woman what she ought to do, so my emotions are somewhat mixed as I attempt it.

Here I should like to distinguish between those women who seek to participate personally in the scientific and technical developments related to the Space Age, and those who, by force of circumstance or otherwise, must make a less direct contribution, which may nevertheless be equally valuable.

The honored maxim about woman's place in the home may sound a little strange at a time when we hear, see and read of interplanetary voyages. Yet if ever there was a period when the affection, inspiration and guidance of a mother could help her children's development, it is the troubled age in which we live. Education must prepare youth for the kind of society which will exist 10 or 15 years hence; we must expect the educational process to anticipate and provide the requirements of the future. Those requirements are clearly definable now, for the impact of science and technology requires at least a fundamental understanding of mathematics and the physical sciences by all our citizens. Not all boys and girls are destined to become engineers and scientists, but if they are to understand the environment in which they will live, they must have an appreciation for the natural laws which will influence their existence more than ever before. Logically, the school should be relieved of additional burdens not related to academic preparation. The place to put them is back into the American home, which must be restored to its rightful position.

It is to the home that we must look to instill in young minds a respect for the basic facts of existence—respect for truth, for initiative, for knowledge, for the satisfaction and reward which can only be obtained from the individual's effort. Today's school teacher occupies a position second only to motherhood, in my opinion, in potential importance to the future of our nation.

For the young woman caught up in the enthusiasm with which youth has embraced rocketry and the exciting prospect of outer space exploration, who wants to share in this intriguing activity, there are many opportunities for participation. One of every four persons employed in the Army's missile and space projects is a woman. Many of them are truly pioneers, in every sense of the word. Some have professional and scientific backgrounds and are working as mathematicians, physicists, chemists, accountants, laboratory assistants and similar occupations. Some actually plot trajectories for satellites and deep space probes. Others calculate the anticipated and actual flight performances of ballistic missiles.

As we regard the future, we must anticipate that the kaleidoscopic happenings of recent years, caused in no small measure by technology, will continue to increase the confusion out of which you are trying to derive order. The first ventures into regions beyond the sensible atmosphere have produced new knowledge which points up the beauty and order of creation, which verifies the natural laws governing all life, and which logically should enhance our reverence. Man will need the understanding help of woman in the intelligent interpretation of these and other revelations which follow in due course as evidences of God's immense plan.

*** END ***

— CHAPTER 10 —

One of the most important convocations on the University of Notre Dame calendar was held annually on Washington's Birthday. This convocation was a tradition that dated back almost to the founding of Notre Dame. To mark the occasion in a special way and to give it added meaning and significance, they established the Patriotism Award; it was presented each year at the Washington Day exercises to "the outstanding patriot of the year who exemplified the American ideals of justice, personal integrity and service to country." The Senior Class chose von Braun as its 1959 recipient ... four years after he became a citizen of the United States.

PATRIOTISM IN THE SPACE AGE [53]
NOTRE DAME 20 FEBRUARY 1959

MR. HAYWARD, REVEREND FATHER JOYCE, DISTINGUISHED GUESTS, MEMBERS OF the University Faculty, and gentlemen:

My emotions are somewhat mixed at this moment. To be honored by one of the nation's great universities, whose 27,000 living alumni have carried the inspiration of Notre Dame into the far corners of Earth, is cause for intense personal pride. But I am also filled with humility and gratitude, because this honor has been given to one who became a citizen of the United States of America only four short years ago.

When I took the oath of allegiance in April 1955, it climaxed a personal ambition for which the seed was laid some 20 years earlier. My brother had spent some time in the Midwest as an exchange student. When he returned from that visit he spoke so persuasively about the warm friendliness of the people, about the energy and progress of American industry, and about the cordial reception accorded to scientific and technological developments, that I then and there developed a strong desire to get acquainted with America as quickly as circumstances permitted. Fate, in the form of a power mad dictator, intervened to postpone my arrival until 1945. It is with a deep sense of obligation that I acknowledge our everlasting debt for the understanding and encouragement which have greeted us everywhere in the United States.

But it was more than the mere fact that we, the recent enemies, were received without resentment that generated an ever-growing desire on our part to be accepted as citizens of this great country. We had left war-torn Germany when it lay crumbled at the feet of the victorious Allies. We were well aware of the fact that by submitting to the rule of a ruthless dictator who had demonstrated time and again that he did not feel obliged to live by the accepted laws of Christian ethics, Germany had burdened itself with an immense guilt.

I consider it the highest accolade which can be granted to any man that he is recognized as a patriot—one who loves his country and who zealously supports its authority and interests. That is especially true in the case of America, because here an entire nation, dedicated to peace, has become an inspiration to free men everywhere by reason of unselfishness and its devotion to the cause of good.

I sometimes wonder how many of our citizens appreciate that to be a patriot in this democracy is much more demanding than in a totalitarian environment. Not only does one's physical situation differ radically, but the moral implications and the responsibilities which rest upon the individual are more complex and are not always appreciated by those born to enjoy the benefits of our society—at least not as keenly as those who, like myself, embraced American citizenship through the process of naturalization.

In a totalitarian state, where the educational process has been subverted to the state's purposes, and where thought control pervades all human communication, there is room neither for question or doubt upon any matter of government authority or interests. Whatever the state decrees, whatever the state does, must be accepted as right and proper. The individual has no responsibility to influence a course of action; in fact, he cannot participate except by the submissive obedience which is always expected of him. He has been thoroughly indoctrinated in the peculiar brand of twisted logic to which dictatorships are addicted, and which is utilized as an effective dose of preventive medicine. He is prepared, therefore, to rationalize and explain, should this become necessary, to quiet his private misgivings or to persuade a fellow countryman who might, by odd chance, retain some spark of individuality. In this exercise he will find excuse for anything which his government does. Some regimes permit him the franchise, but only under rigorous controls and restrictions which allow him no freedom of choice, no opportunity for self-expression. His ballot becomes little more than an automatic endorsement, which is nevertheless held up to the world as an example of the democratic process, distorted beyond recognition.

To the native-born American, to whom all of this is an ugly nightmare because he need never experience it, it must seem incredible that anyone living under these circumstances could accept them without resistance. He has been educated in the democratic tradition. There are readily available to him the means and methods by which to influence local, national, even international policies. The incredible fact is that he is freed to [express himself] without courting arrest or disfavor. It is expected of him.

If, however, he refuses to exercise his prerogatives, if he turns his back on the shaping of policy and the fulfillment of government programs, he disqualifies himself [from] the role of patriot. Inherent in that term is the willing assumption by the individual of the fundamental obligations of citizenship: to keep himself informed; to apply his best judgment in the light of the facts and circumstances; and to act in accord with his convictions and the dictates of his conscience to the end that the best interests of his country are served. I propose now to explore my contention that this is indeed a time when every American should understand the threat and when he should gladly accept the duties of patriot, however onerous, however unpleasant they may be.

I suppose that most of us at some time or other have been dismayed by the apparent lack of zealous interest by the public with respect to critical situations. Our first president, George Washington, commented nearly 175 years ago on the tardy operations of democracy in a letter to his good friend, Marquis Lafayette. "It is to be regretted, I confess," Washington observed, "that democratic states must always *feel* before they can *see*; it is this that makes their government slow, but the people will be right at last."

History has borne out Washington's judgment in the great crises which the nation has survived; the first and second World Wars and the Korean Conflict are outstanding examples. Once aroused, America's might exercised on the side of right has prevailed against all dangers. In this age of ballistic missiles and satellites we may no longer be able to repeat what was accomplished in 1917, in 1941, and even as late as 1953. Unprecedented new technologies have practically erased the protective rampart of time and space behind which, in the past, this nation could mobilize its resources. Reaction, if it is to be at all effective, must now be well-nigh instantaneous, and time for preparation will no longer be available.

The Soviets have achieved downright alarming progress in rocketry. The Soviets have demonstrated beyond a shadow of a doubt that their educational system can produce brilliant scientists and engineers who can do far more than copy Western technological advances, men who can successfully pioneer in areas which no one has yet treaded upon. The sober fact is that these men managed to steal a march on us.

There is another side which should be emphasized and that is the purpose to which we will devote our space program. It must not be to achieve parity and then supremacy in a weight-lifting contest alone. For the Communist challenge extends into many areas—cultural, economic, educational as well as technological and political. The test is not confined to the comparative abilities of rocket development teams—it involves every man and woman in America.

The United States must increase its resources in trained manpower by training more scientists and engineers in its atmosphere of academic freedom. The men who founded Notre Dame anticipated the course of progress by introducing a scientific curriculum in 1865 and by adding the engineering curriculum in 1873. If we are to continue to enjoy an ever-rising standard of living we must encourage and nurture technology.

Every citizen who is to feel at home in the kind of environment which technology will create a few years hence must possess a broad and basic understanding of the physical sciences. Knowledge of this kind will be fundamental to the art of living if man is to control the machines he creates and the forces available to him—forces which can either devastate the Earth or can be wisely employed in a golden age of opportunity and adventure, in which he will soar into the far reaches of the Universe.

In his farewell address, General Washington urged his countrymen to "promote as an object of primary importance institutions for the general diffusion of knowledge ... in proportion as the structure of a government gives force to public opinion, it is essential that public opinion should be enlightened."

It behooves every man who considers himself a patriot to heed that admonition, and to do whatever he can to further the cause of enlightenment to the end that our people should be fully aware of the demands of the future. If they truly "feel" and "see," they will arise to the need as they have met all the other needs in their progress towards a better life and a peaceful world.

*** END ***

— Chapter II —

Lieutenant Fred Kleis was assigned to von Braun as a military aide. The young Lieutenant was unreservedly at his "beck and call." As an indication of their mutual respect, von Braun took time from his busy schedule to speak to Kleis' alma mater, Pennsylvania Military College,[54] Chester, Pennsylvania on June 1, 1959.

THE OPPORTUNITY TO SPEAK TO YOU TODAY IS A PLEASURE, FOR I HEARD MUCH about Pennsylvania Military College (PMC) in the past. And since I accepted the invitation to be here today, I have heard practically nothing but the merits of PMC from my aide, Lieutenant Fred Kleis. He is a graduate of the class of '54.

Fred has told me of your school's revered traditions and historical past, its influence through the years, the excellence of your cadet program, the superiority of your professors, and reeled off a long list of students of former years whose names are familiar to everyone, such as Cecile B. deMille and Bill Stern. In fact, he so thoroughly brainwashed me, that when the "Old Main" came into view this morning, I felt like a homecoming alumnus myself.

Although the information he gave me came from what might be considered a prejudiced source, I felt certain that it is 100 percent correct. A school which can instill such loyalty in the heart of its graduates is surely deserving. All of you may well be proud of the many notable figures who have gone from these halls of learning to carve out distinguished careers in military and civilian fields as engineers, statesmen, industrialists, authors, and officers.

I would like to direct my remarks primarily to the students who are today entering these highly competitive career fields. You have made excellent preparations through your choice of school, in which you could combine the mental and physical disciplines of military training with your academic studies. You will need both to cope with the complex problems of tomorrow's world.

The swift progress of scientific discovery and technological exploitation has already created an environment in which change is the dominant force. As its accelerated pace increases, unpredictable events will test human faith and fortitude, and dynamic evolution will gravely influence the stability of social order. Only by the exercise of wisdom, restraint, and with a high sense of moral responsibility can man wisely control the energies available to him as he enters the Space Age. In a day when hand-picked astronauts are being trained to explore outer space, when we are planning expeditions to the Moon and nearby planets, a new scale of values demands our earnest consideration.

Young minds appreciate the true meaning of the big missiles we are developing. To you they are glorious transportation systems—not simply means of destruction. They are vehicles for carrying men or cargo between points on Earth, or points in space. They are the means of exploration for the Marco Polos and Magellans of tomorrow.

It is little wonder that some impatient youths want to leapfrog the difficulties of mastering physics, chemistry, algebra or calculus, and take off instantly to the Moon. I can assure you from personal experience, that it is better to pay attention and get the fundamentals well in hand as early in life as possible. I cannot help pondering what might come out of our schools if we spent as much effort and money to develop students as we do athletes.

I suspect that we have not fully awakened to the breadth and depth of the Soviet challenge that today faces men who love freedom, men led in their efforts to avoid enslavement and seek justice and peace by the United States. I cannot emphasize too strongly that the Soviet challenge is by no means restricted to military technology. It goes far beyond the realms of politics and armies, and involves every facet of our civilization, every segment of society.

I know the price of lost freedom, having survived one dictatorship and enjoyed the privileges of American citizenship and residence in this country. I don't want my children or yours to pay that price because of our failure to recognize the seriousness of the threat that confronts us as we enter the Age of Space. This can be the most fruitful and most enjoyable of all the periods of recorded history, or it can end in catastrophe so awful as to defy description. My concern that we make the right choice has kept me away from my family and my job week after week during the past two years, whenever I have had the opportunity of addressing thinking groups of people on the dangers, opportunities, and needs of these critical times.

There can be no possibility of mistake in evaluating Soviet intentions so far as outer space is concerned. Their objective is to master the spatial environment, for scientific purposes initially, but with the long-range goal of putting the capability of manned travel through outer space to use for the same overriding goal of expansion of their sphere of influence which is the ultimate purpose of all Communist endeavor.

I am convinced that it is man's destiny to enter space and that he who controls the open spaces around the Earth is in a position to dominate its peoples. The United States was momentarily jarred from its complacent belief that we held monopoly on technological advances by the launching of *Sputnik* eighteen months ago. With suddenly increased emphasis we successfully launched a few satellites and deep space probes, and recovered the first primates from flights into space.

After each advance I am besieged with the questions: "Have we caught up with the Russians?" and "Who's ahead now?" These people say, "The Russians haven't launched anything lately; they must be having trouble with their programs." We may rest assured that they are not standing around with their hands in their pockets. I would suggest that we brace ourselves for other Soviet firsts in the new field of astronautics.

A large segment of our population seems to take for granted American supremacy in all human endeavors, and assume that whenever we wish to do so, we can overtake the Russians. This can be downright dangerous; at best, it is foolhardy. At the same time, I disagree with alarmists who cry that we have lost our position as a first-rate power. That kind of sweeping generality can result in hysterical, ill-conceived reaction which can be wasteful and harmful.

The Soviet effort in space antedated that of the United States. They felt the need for intercontinental ballistic missiles, began work on this particular weapons system earlier than the U.S., and coupled their military advances with scientific space exploration from the beginning. In the Soviet Union the decisions are made by a few men on the goals toward which the country's resources and the labor of its people shall be directed, and how fast this can be accomplished in the best interests of their Communist dream of expansion. In this country we can muster the necessary resources for a plan best suited to our objectives only if the people themselves want it, will steadfastly support it, and are ready to pay for it. This may well become our last chance for such a determination. In the past, Americans have always reacted adequately, after becoming aware of the need, and marshaled its resources.

Money alone will not turn all the tricks, although there are times when lack of it causes a project to be delayed or canceled entirely. It would be presumptuous for me to offer any single solution, or even to list all the things that must be considered. But there are some minimum requirements which can be identified and which demand prompt action. Not the least of these is the educational program, a subject of such sensitive nature that an "outsider" like myself approaches it with trepidation.

If a speaker applies the broad brush of approval in talking about schools, he may delight his audience at the cost of his integrity. If he attempts anything like a critical appraisal, he risks displeasure or worse. Yet it should be entirely logical to expect a keen interest in education from an informed citizenry. In the field of scientific and technological education the Soviet Union has been eminently successful. Their state-controlled educational system is turning out competent engineers and scientists in greater numbers than ours.

Military training is like running a nursery, where the crooked saplings are weeded out so the sound and straight ones can survive and grow stronger and bigger. When you think, however, of the mental tortures and the flashy temporary successes followed by breathtaking setbacks which mark the life of an artist or a scientific pioneer, the value of this kind of training becomes questionable. The creative artist or the scientific genius might be likened to the orchid, which flourishes best in the jungles, not the nursery. I've never considered myself a genius—and my wife is always ready to attest to this fact—but I assure you I would have flunked most of those exams in a survival-of-the-fittest system during the tender years of 16 through 22.

This method removes from positions of possible productivity those who fail exams, which may not truly test their creativity. It does select students who are apparently capable of absorbing advanced knowledge, and underlines the need for higher quality educational product, rather than a quantity of mediocre talent. Dr. Harry Gideonse, President of Brooklyn College, has said that the great challenge of 1959 calls for training of our best minds on all levels to their full potential. I am sure there are some teachers, not in this College of course, who agree with George Gissing that "education is a thing of which only the few are capable; teach as you will, only a small percentage will profit by your most zealous energy." A good case can be made out to support the arguments that we shape our educational offerings too much in terms of mass and not enough in terms of quality, or in terms of the individual's peculiar requirements. I think the mass approach is just and necessary, that the individual who possesses even a faint spark of creativity, may have it fanned into full flames. But once he catches fire, the educational system should be geared to open the doorways to man's total accumulated knowledge, rather than smother him with lack of fuel or oxygen.

Material computation does not decide the destiny of mankind, as Winston Churchill observed. When great causes are on the move in the world, we learn that we are spirits, not animals, and that something is going on in space and time which, whether we like it or not, spells duty. In the same context, I want to leave with you a statement by Thomas Huxley: "Perhaps the most valuable result of all education is the ability to make yourself do the thing you have to do, when it ought to be done, whether you like it or not. It is the first lesson that ought to be learned; and however early a man's training begins, it is probably the last lesson he learns thoroughly."

I might add that the day he learns it, boyhood ends and manhood begins.

The sense of duty must be supplemented by other qualities as you advance your careers. One of these is initiative. The drive towards accomplishment distinguishes one man from his fellows, because it is he who exerts that extra measure of effort which separates him from the group. He ventures into the unknown; he takes the calculated risk in seeking a worthwhile objective. This is the force propelling us into the future. Every leader has a large measure of initiative; without it, he will not long retain leadership.

Closely related to initiative is the faculty of imagination. Thomas Edison commented that invention is 99 percent perspiration and one percent inspiration. His point was that hard work is necessary to success. But the corollary is that without the one percent inspiration, all the other effort would be either wasted or repetitive. Never before has there been so acute a need for men of vision when the challenge is as limitless as space itself.

To support initiative and vision, there must be moral and physical stamina, the ability to resist fatigue, to cling tenaciously to what you consider the right course, and to reject the easy and obvious answer. Some of life's knottiest problems involve the choice between principle and compromise. Often the latter appears to be expedient; often there is no other way out, but no one can continue to reject absolute truth without risking disaster. Between the choice of what is popular and what is right, there can be no time for hesitation.

Somewhere, too, in the learning process every man must understand the value of respect. The trained mind respects the lessons of history, the immense body of truth embraced in the sciences and humanities, the institutions of law and government, and above all, the beauty and order which are characteristic of the universe outside the sensible atmosphere.

Our first steps into the vastness of space have served to demonstrate more emphatically the law and order of the universe, which we behold with awe and wonder. Everything which we have found there serves to enlarge our concept of the powers of the Creator, and strengthen our faith in the conviction that a kind providence guides our steps. An unshaken faith can be your strongest shield against the stresses and strains to which you will be subjected as science comes to maturity. Faith has been the solid foundation on which this nation has grown to greatness. Holding firm to this heritage, you can look to tomorrow with courage and confidence that by your progress all men of good will can live in peace.

*** END ***

– Chapter 12 –

Greater Boston Chamber of Commerce [55]

The Greater Boston Chamber of Commerce invited von Braun to accept the Boston community's first Distinguished Leadership Award in recognition of his outstanding contribution to America in the missile age, May 1959. The Honorable John H. McCormack placed von Braun's speech in the Congressional Record of the 86th Congress with the following remarks:

"On May 27, 1959, Wernher von Braun, Army Ballistic Missile Agency, U.S. Army Ordnance Missile Command, Redstone Arsenal, Alabama, gave a very important and significant address to the members of the Greater Boston Chamber of Commerce, Boston Massachusetts. In his address, Dr. von Braun said, among other things:

> 'We are confronted by an aggressive competitor whose ultimate objective requires that he exploit his advances to counter and defeat our programs, not only in science, but in every phase of our social order. Those of us who are engaged in missile and space programs feel this challenge keenly.
>
> 'It would be utterly fallacious to assume catching the Russians in the race for space will rid us of Ivan's shadow. The Soviet challenge is political, ideological, cultural, economic and educational—it is as broad as human society and social organization. For Communism aims at world domination and nothing short of that.
>
> 'I believe we must awake America to the danger and then, as a united country, get to work.'

"He then referred to his own experiences under a dictatorship, when Dr. von Braun said:

> 'Having lived under a dictatorship equally as cruel, equally as ambitious as is the Communist state, I know the price of lost freedom. I don't want my children or yours to pay it because we failed to recognize the real threat in all its proportions.'"

McCormack stated, "In my extension of remarks I include the excellent and dynamic address, to which Americans should pay heed, made by Dr. von Braun on May 27, 1959."

A title was not assigned to the following speech given in Boston.

BOSTON HAS BEEN CALLED THE "HUB OF THE UNIVERSE." IT WOULD THEREFORE be logical, it would seem to me, that the citizens of this Capital of the Solar System should have more than a passing interest in the activity that has taken place in those regions during the past two years. The officials and citizens of this Solar Capital must know what is going on in the wards and precincts if they justify their title as the "Hub," and in this case the wards and precincts are in Outer Space.

We have another center of space activity in the South that would like to share with you this honor. It is the headquarters of the Army Ordnance Missile Command at Huntsville, Alabama, or better known in those parts as "Rocket City, USA." There the missile systems were designed and fabricated which launched the Explorer Earth satellite January 31st, 1958, the first launching of its kind in the Free World. There the Juno II rocket was built which launched the *Pioneer IV* space probe March 3, 1959—the probe which passed the Moon and went on to become the first made-in-the-USA satellite of the Sun.

When I was invited to speak in Boston, my first thought was of the great universities in and around it, and of their great teachers and splendid scholars. So naturally, confronted and surrounded as I am here in Boston by such highly educated fellow citizens, and inadequately prepared, I am comfortably reminded of the abiding truth of those words that never occurred to Horace, but which were uttered by Adlai Stevenson in his Godkin Lecture at Harvard: "*Via ovicipitum dura est,*" or, for the benefit of my fellow engineers: "The way of the egghead is hard."

When we attempt to forecast the dynamic progress of technology, the problem is equivalent to measuring a turbulent river whose volume increases geometrically by incessant rainfall. Scientific research is a self-generating and self-multiplying phenomenon. A new discovery contributes to vertical building of the scientific structure. It also brings with it opportunities for still more advances. Each new discovery in fundamental or basic science opens up challenges for the applied scientist and the engineer to interpret the results, to devise means of utilizing them in new products, which enrich our standard of living. Science and technology create the momentum which carries them forward at [an] ever-accelerating rate. Consequently, it is extremely difficult to predict in an absolute sense the direction in which developments may occur, the kind of end products which may stem from them, or the magnitude of the industrial effort they may spawn.

Practically everything we utilize is the result of research and development in the past century. Every modern convenience derived from some type of research, usually performed without regard to its application, by men who simply wanted to know more about the universe. This was true of the automobile, the airplane, the clothes we wear, the food we eat, the electrical miracles which brighten our homes.

One major area of technological advance which has caught the imagination of mankind is, of course, the development of rocket propulsion and guidance systems which will enable man to explore the limitless reaches of outer space. The field is moving at such a pace that the danger is in underestimating the possibilities.

A second major area, dating back approximately 20 years, is that of nuclear energy. Since the first atomic bomb was detonated a little over 13 years ago, many developments have taken place. Within the [last] year, the first American commercial generating station powered by nuclear energy has been put into operation. This should be regarded as the forerunner to the use of power systems which tap the vast resources available in the radioactive process. We have seen nuclear energy applied to the propulsion of submarines and we can expect increasing use both of the direct application of nuclear energy and the indirect utilization of many of its byproducts.

Still a third area is automation, the application of machines to control other machines and so to achieve automatic functioning. In its simplest form automation has been applied in the mass production techniques of the automotive industry. More recently it was introduced in the electronics industry where numerous assemblies utilized in radio, television and other devices are automatically assembled, tested, packaged and shipped.

These are but three of the numerous areas of technological advance, but will suffice to emphasize its far-reaching influence and to point up the unpredictability of the scientific future.

Now I will attempt to relate the technological and the human equations, because of my sincere belief that we are so mesmerized by our successes we may overlook, and therefore fail to prepare, for the satisfactory resolution of human problems generated by our progress. If we forget man in our enthusiasm for science and technology, inspired by their awesome achievements, we shall not long survive to enjoy them.

One of the major differences between the demands of space, nuclear and automation industries from our industrial environment of the past lies in the quality of labor force that is needed. The textile industry I cited earlier required labor capable of learning certain routine operations. Very little understanding of principles was involved. With the introduction of automation, more and more routine processes will be performed by machines. These machines need to be controlled and maintained. Their operators must be prepared by a higher level of training—the requirement is [for] more technically and scientifically educated men at the lower levels of [the] industrial strata.

Another difference will be found in the kind of scientists and engineers required. There is evil in overspecialization; there is basic need for broad, general scientific training. The man trained in a narrow scientific or engineering field may find himself left behind by the technological developments of tomorrow. Flexibility has become the must in professional preparation. We shall require increasing numbers of men thoroughly trained in the fundamentals of science so that they can adequately pursue the developments of the future and adapt themselves to new and unforeseen requirements as they arise.

As automation replaces unskilled labor, the productivity capacity of the individual will expand. Inevitably this will lead to a shorter work week and more leisure time. We must assure that leisure time does not become wasted time, a trend that is all too apparent at present. Almost every individual who is to feel at home in the world, who is to function as an intelligent member of society, will in the coming years require a broad basic understanding of the physical sciences as surely as the adult today must know how to drive an automobile and twist the dials on a TV set. This knowledge will be fundamental to the art of technical living. We must reach out to the largest possible segment of our population and create an understanding of literature, philosophy, history and sciences which form the foundations of our culture.

I believe there is a great challenge in this situation confronting the people of Boston. If, against these readily foreseeable requirements, we pause to take stock and determine what we are doing now and its relation to what will be required of us, I suspect that we will discover some problems which should be attacked with a sense of genuine urgency. By defining the problem, we can set about developing a solution before it becomes overpowering. There is an equally valid, perhaps even more pressing reason for action now. We do not enjoy a monopoly on technology. We are confronted by an aggressive competitor whose ultimate objective requires that he exploit his advances to counter and defeat our progress, not only in science, but in every phase of our social order.

Those of us who are engaged in missile and space programs feel this challenge keenly. The implications have not been thoroughly grasped by all of our people if we are to judge by the apparent disinterest they manifest. Evidently they are happily convinced that we are supreme in all fields and will assuredly overtake the Russians in their bid to exploit outer space for the purposes of Communism. We enjoy superiority in many areas, to be sure, but we must ask how long that situation will continue and look into those areas where the assumed superiority does not exist.

It would be utterly fallacious to assume that catching the Russians in the race for space will rid us of Ivan's shadow. The Soviet challenge is political, ideological, cultural, economic and educational—it is as broad as human society and social organization. For Communism aims at world domination and nothing short of that.

The threat has become all the more dangerous because of the impressive technological capability built up by the Soviet state through a massive educational program designed to provide an immense reservoir of scientific and engineering skills. While the long-range purpose of the Russian space drive may be shrewdly masked under the guise of scientific effort, I believe they intend to expand their sphere of influence in every direction. If we do not aggressively pursue an equally determined United States space flight program, with an even greater sense of urgency, we may in the not to distant future find ourselves surrounded by several planets flying the hammer and sickle flag.

How much time remains is indeed a moot question. It is not the scientist whose research paves the way, or the engineer who translates his discovery into useful hardware who decides when, how and for what purpose space will be exploited. Those decisions and their realization depend primarily upon the level of effort committed to the tasks. I can translate that into how many dollars and cents are made available to carry on the work. In the Soviet Union those decisions are made by a few men concerned only with their dreams of expanding Communist influence. In this country we can muster the resources only if the people themselves want such a program, will support it over whatever period of time is required, and will sacrifice if necessary to achieve the desired ends.

I believe we must awake America to the danger and then, as a united country, get to work! Having lived under a dictatorship equally as cruel, equally as ambitious as is the Communist state, I know the price of lost freedom. I don't want my children or yours to pay it because we failed to recognize the real threat in all its proportions.

*** END ***

– Chapter 13 –

From 1930 to 1941, Robert Hutchings Goddard conducted his research in the development of high altitude rockets at his shop at the Mescalero Ranch and in Eden Valley just northwest of Roswell, New Mexico. Goddard's investigation covered almost every essential principle involved in the theory and the actual practice of high powered rockets of the day. The formal opening of the Robert Hutchings Goddard rocket collection was held on April 25, 1959 and von Braun was invited to speak at the dedication. Esther Goddard wrote von Braun expressing her thanks for his kind remarks about her husband. "I was almost overcome with the great kindness of your remarks about my husband, showing a generous heart and a broad understanding. Mr. Hersey has told me that this talk is to be printed in *Astronautics*. Please accept my warmest thanks."

Dedication of Goddard Roswell Museum [56]
15 April 1959

MR. ASHTON, GOVERNOR BURROUGHS, SECRETARY MILTON, MRS. GODDARD, distinguished guests, ladies and gentlemen:

I am honored to participate in the dedication of the Robert Hutchings Goddard Rocket Collection and the Paul Horgan Gallery of Roswell Museum and Art Center. I am doubly honored to be invited to discuss the contributions of my boyhood hero, Dr. Goddard, to science, although this is indeed difficult because the towering importance of his pioneering efforts is impossible to evaluate at this stage of the art.

However belated, the recognition accorded to Dr. Goddard is eminently deserved. Like other scientists working in novel and little understood fields, venturing into unknowns fraught with hazard, he did not live to see the fruits of his successful work. Now he belongs to the world of space.

Rocketry in this country and abroad owes a great deal to his vision. While he contributed basic knowledge essential to further progress in his field, he also demonstrated an extraordinary ability to translate theory into operating rocket systems. In the light of what has happened since his untimely death, we can only wonder what might have been if America realized earlier the implications of his work. I have not the slightest doubt that the United States today would enjoy unchallenged leadership in space exploration had adequate support and recognition been provided to him. It is to the great credit of both institutions that the Guggenheim Foundation and the Smithsonian Institute gave at least a part of his continuing effort some financial backing and appropriate professional recognition.

He was engaged in studies of high altitude research rocket design while still a student in 1908. In 1919 he published a monograph titled *A Method of Reaching Extreme Altitudes*. On March 16, 1926, he launched the world's first successful liquid fueled rocket in Auburn, Massachusetts. Three years later, one of his rockets transported an instrument package that housed a barometer, thermometer and camera. With the understanding aid of General Charles Lindbergh, and the Guggenheim Foundation, he established a modest research facility at Mescalero Ranch in Roswell and that historic association gives full validity to the selection of this site for the permanent record of his work.

On April 19, 1932, he demonstrated an automatic gyroscopic control mechanism. On March 28, 1935 he flew his first gyro-controlled rocket, and on May 31st that year, one of his rockets soared to an altitude of 7,800 feet. In 1936 the Smithsonian published the last of his reports for that institution, a record of his success in liquid propellant rocket development. He was the first to employ vanes mounted in the stream of escaping gas for the purpose of steering the rocket. He patented the concept of the multi-stage rocket, the principle employed in all our space exploratory efforts to date. He formulated the mathematical theory of rocket propulsion and flight, and first proved mathematically that rocket propulsion will function in a vacuum. He invented the first smokeless powder rocket, and was the first to recover instrument containers from rockets by parachute technique.

The American Rocket Society has long since recognized his courage in sticking to his beliefs in the face of skepticism and indifference, heart-breaking technical difficulties, and with meager resources. The Society has credited him with developing rocketry "from a vague dream to one of the most significant branches of modern engineering."

In his life and work Dr. Goddard was the embodiment of a truth which is frequently overlooked in our characteristic reliance upon team effort. Of course, Goddard too was supported by a competent team, and most of its members are here with us today. But for all the credit so richly deserved by this team, it is well to remember that any of the most significant advances in science and technology have not emanated from teams but from the singular efforts of dedicated and solitary individuals.

The formula to any success in life is the exercise of individual ability, and the formula for outstanding achievement in science, industry or the arts is single-minded devotion. This is the one quality which distinguishes the truly great man, like Goddard, from his fellow citizen. Since time immemorial it has been the single-minded devotion of dedicated men which has provided the human race with immortal works of art, with great books, great teachings and great discoveries.

The late Thomas Alva Edison regularly locked himself in a laboratory for days on end. He, too, had that rare capability of single-mindedness which allowed him to put every other consideration out of mind and concentrate entirely upon the problem at hand.

Albert Einstein had that same quality of genius. One day he stopped a lady on the street in Princeton, where he lived, and asked if she could tell him where he was. "Your home is right across the street," she replied. "I did not notice that," Einstein commented.

Many of the world's leading thinkers and doers have been considered "eccentric" by others' standards. The significant thing is that they did not care, or that caring, they nevertheless had the fortitude and the conviction to carry them on. In spite of all drawbacks and discouragements they persisted in their chosen work, impelled by a secret fire which drove them on quite oblivious to the wishes or expectations of society. I suspect they would be labeled "visionary dreamers" in these days. Let us not forget that it has been the "visionary dreamers" who have accomplished more for the edification and benefit of mankind than any other contributor to civilizations past and present.

If, as a people, we ever lose the capacity to continue exploratory effort in science and technology, with or without social blessing, we shall have stifled our mental rebels. Yet it is the mental rebel, the incorrigible non-conformist, who pursues his convictions despite all mockery or indifference, who is God's most precious gift. Without men ready and capable of thinking bold, new thoughts, of challenging the accepted theorems, the progressive development of human society would grind to a halt.

Fifty-two years ago and not yet a doctor, Robert Goddard submitted an article suggesting that atomic energy would one day propel a rocket into interplanetary space. A magazine editor declined the paper with thanks and told him that "the speculation is interesting, but the impossibility of ever doing it is so certain that it is not practically useful. You have written well and clearly, but not helpfully to science as I see it."

Practicality—what wonders have you denied mankind!

Perhaps there is the abiding lesson of Robert Goddard's life, the common insistence of the oh so practical men upon measuring his proposals in terms of concrete, measurable, immediate results, the failure to understand his penetrating visions, to accept his revolutionary concepts.

The creed of the "practical" should go something like this:

> What I cannot see cannot exist.
> What I cannot touch is not there.
> What I cannot hear I do not understand.
> What I cannot taste has no substance.
> What I cannot smell is odorless.
> What I cannot understand I do not believe.

There is the urgent challenge of the time as we venture across the threshold of limitless space, a challenge fresh in our thoughts on this occasion as we pay tribute to the man who dared. We must open our vision to the unknown; we must expect the unpredictable; we must value knowledge for its own worth; and we must cease to measure the new in terms of usefulness alone.

Dr. Goddard's contributions have provided a solid basis for the progressive development of rocketry as the means by which to achieve his shining ambition—the exploration of the space regions where the silent planets, stars and galaxies await the adventurers who follow in his giant footsteps.

*** END ***

– Chapter 14 –

University of Florida [57]

MY TRIPS TO FLORIDA, FOR BUSINESS AND PLEASURE, HAVE BECOME SO FREQUENT that I almost consider your state my second home. This is not because of the length of my stay each time, for that is all too brief. But it is because of the moments of pleasure, relaxation, and sometimes even pride and exaltation crammed into visits of a few hours.

Most of my flying trips are to Florida and Cape Canaveral, where our ballistic missiles and space vehicles are launched. The events of several early morning hours spent in a concrete blockhouse on that isolated sand pit are indelibly impressed in my memory. One of these nights was March 3 of this year, when *Pioneer IV* was started on its journey into space.

I would like to explore with you some of the obvious implications in our present position in space exploration with relation to that of an aggressive competitor determined to conquer space for his own ends. I would emphasize at the outset that the Soviet challenge is by no means restricted to military technology. It goes far beyond the realms of politics and armies. The task of coping with the Red menace to our security and our future is no longer the exclusive responsibility of generals and statesmen. The struggle involves every facet of our civilization, every part of our society: religion, economics, politics, science, technology, industry and education.

I know the price of lost freedom. I don't want my children or yours to pay it because of our failure to recognize the seriousness of the threat as we enter as Age of Space, which can be the most fruitful and the most enjoyable of all the periods of recorded history, or which can end in catastrophe so awful as to defy description.

I would suggest we brace ourselves for Soviet "firsts" in the new field of astronautics. We are behind and we cannot catch up in a day or two since major technological projects necessarily involve lead time. It will require several years of concentrated effort to come abreast, and even longer to pull ahead.

We can waste no time commiserating over the sorry lot of the Russian worker or peasant, comparing his lack of freedom and creature comforts to our prosperity. We should also shuck off another illusion, that the Russian people will rise up to overthrow the Kremlin and thus relieve us of all worry.

Perhaps a dream of freedom exists somewhere in the Soviet Union. Perhaps by exposing young minds to scientific training, the search for truth will eventually lead to unmasking the dictatorship. But we cannot stand around, hands in pockets, waiting for others to do what can only be accomplished by us. I am convinced it is man's destiny to enter space and that he who controls the open spaces around the Earth is in a position to dominate its peoples. Our choice is to accept the Soviet challenge or pay the piper.

If we do not match the ambitious Communist intentions to visit the Moon and other planets with an equally determined United States space flight program, pursued with a real sense of urgency, we may in the not-too-distant future be surrounded by several planets flying the hammer and sickle flag.

In the field of scientific and technological education the Soviet Union has been eminently successful. No small part of this success is due to the fact that Russia has, by our standards, a "single channel career society." Let me elaborate a bit. The young Russian who wants to become an engineer must pass a series of examinations before admission to an engineering college or, if his aspirations run all the way to the top, to the Moscow Institute of Technology. As all education in the Soviet Union is free, a boy cannot buy his way into any of these schools, although it may be presumed it would be helpful if Daddy were some kind of a Big Wheel. Passing the entrance exam is the necessary first step. But it is only the beginning.

Year after year the Russian student is subjected to new examinations designed to wash out the less competent and to arrive at the kind of educational pyramid the Soviet government needs. Year after year the student faces the possibility of washing out. If he fails to acquire a bachelor's degree in engineering, he is shunted off for life into the career of a draftsman. If he passes the bachelor's requirements but fails the master's degree, he may become a layout designer or a shop superintendent, but that must be the end of his ambition. He must acquire the master's degree, or even better, a doctorate in engineering if he expects to reach the summit and make a personal dent in technological progress.

What is the moral of all this? Well, the Soviet Union does not have anything to offer which compares with the career of a successful U.S. businessman. In Russia you either pass those examinations or you've had it, and you are channeled for life into a lower strata of Soviet society, with practically no hope of recovery. This is what I mean by the Soviet Union's "single channel career society."

More than ever before our schools, colleges and universities have become the bulwark of our cherished freedoms. We cannot afford to have scientists and engineers who, in their fields, are inferior to their Soviet counterparts. But in order to be sound citizens of a free nation, our knowledge and interests must embrace a vastly larger area and extend well beyond relatively narrow professional occupations. This simply means we must learn and work harder than the men behind the Iron Curtain. Let us ask ourselves if we are really doing that. Let us search our minds and souls to determine if we have faced up to the challenge of the Space Age.

Now, I would not have you think nothing has been done by this country since the Russians made their opening bid with *Sputnik*. A great deal has happened in the last seventeen months. Two new agencies have been established to carry forward our space programs—the Advanced Research Projects Agency (ARPA) of the Department of Defense (DOD), which is concerned with space programs of military significance, and the National Aeronautics and Space Administration (NASA). The objectives of the United States have been stated by President Eisenhower in clear and unmistakable language. When he spoke of the opportunities which a developing technology can provide to extend human knowledge of the Earth, the solar system and the universe, Mr. Eisenhower said, "These opportunities reinforce my conviction that we and other nations have a great responsibility to promote the peaceful use of outer space and to utilize the new knowledge obtainable from space science and technology for the benefit of mankind."

In essence, each of the payloads hurled into space by U.S. rocket systems have been small-sized, scientific laboratories equipped with highly efficient sensing and measuring devices. More ambitious space projects are under way. Let me mention, in passing, one of the less obvious but interesting aspects of our achievements. Just one of the Earth's satellites, *Explorer I*, has traveled nearly 150,000,000 miles since it was injected into orbit January 31, 1958 from the Florida coast.

There are many intriguing possibilities in this space business. For example, a set of three properly spaced communication satellites equipped with modern electronic recording equipment could easily handle the entire mail volume of the world—[a] postal service, acquiring radio messages, as one example, and relaying them via line-of-sight links to New York, Rio de Janeiro, London or Bombay. Worldwide television and broadcasting service can be achieved with the same satellites. They would be positioned above Earth at correct altitude so that they remained in a relatively stationary position. One satellite will be used for weather observation and forecasting will become at last an exact science with its help. The savings in lives and property damage by advance hurricane warning alone could far exceed the cost of the service.

At the outset I emphasized the need of strengthening our claim to leadership. I believe we must begin by preparing our youth adequately to cope with the problems and the contests that lie ahead. We must disabuse ourselves of the idea that school is a place in which boys and girls learn how to live together and nothing else. They must understand mathematics and the physical sciences, and they need more, better-prepared teachers, who can only be attracted to their profession by better salaries, improved professional status, and the kind of physical facilities which encourage inspirational leadership to interest young minds in facts.

We must generate the will to supremacy. Because this is intangible, because it must come from the hearts and minds of our people, it cannot be legislated, budgeted, or evoked by decree. We stop telling the world what we are against. We should tell the world what we are for. We must not fight Communist ideology with negative statements, but with lofty ideals of [the] founders of this great Republic. The antidote to Communism is not anti-Communism, but the belief in God and the dignity of the individual. Let us not deceive ourselves; the Communist ideology has powerful appeal to the have-nots, the uninformed, and the desperate. But ideas are fought not with material means but with superior ideas translated to the benefit of man. Where should these ideas be found in this world if they cannot be found in this glorious Land of the Free?

The rocket developer with his eyes upon the outer galaxies is certainly not motivated by material reward. Yet he is constantly asked, "Why do we do these things?" "Will they ever pay off?" Can you imagine anyone asking that of an Einstein or a Pasteur? There will be payoffs, of course, as we move into space, but the most important, the reward of greatest lasting value, is our increased understanding of the spatial environment and of its influence upon our Earthly existence.

You cannot put a price tag on such a discovery as the Van Allen Radiation Belt, made possible through the launching of relatively small orbital satellites [*Explorer I*]. We have opened the door into the limitless reaches of the universe by these modest beginnings, and we can see just far enough ahead to know that man is at the threshold of a momentous era. Here is opportunity, challenge, adventure so tremendous as to exceed anything which has gone before. Here is the tomorrow which youth wants to embrace and for which we must provide whatever guidance and preparation we can. I am confident that as more of our educational institutions conduct programs like that which is carried on in this campus, we shall begin to prepare for the new age.

*** *END* ***

— Chapter 15 —

Space, Its Problems and You [58]
Presented before the Mississippi Dental Association Biloxi, Mississippi

MEMBERS OF THE MISSISSIPPI DENTAL ASSOCIATION, LADIES AND GENTLEMEN:

When I look out on this audience of men engaged in a profession which alleviates pain, I am reminded of Shakespeare's humorous statement: "For there was never yet a philosopher that could endure the toothache patiently." I am also reminded of another rather lighthearted verse which brings both your and my professions together. It is from Gilbert's *To the Terrestrial Globe*:

> *Roll on, thou ball, roll on!*
> *Through pathless realms of Space*
> *Roll on!*
> *What though I'm in a sorry case?*
> *What though I cannot meet my bills?*
> *What though I suffer toothache's ills?*
> *What though I swallow countless pills?*
> *Never you mind!*
> *Roll on!*
> *(And it rolls on)*

I am going to discuss space, its problems, and you—not simply you gentlemen of the Mississippi Dental Association, but the American people. The ballistic missile enterprise is rapidly assuming such proportions that, if we speak of the total guided missile effort, it may well become the largest enterprise of its sort in the nation's history—at least in peacetime. That is understandable if you realize what we are attempting: nothing short of the exploration of the universe. Here is the last, the greatest frontier of all, so immense that it requires new concepts in mathematics and physics to express it in scientific terms.

During this past fiscal year, which ends June 30, 1959, the missile budget of this country involved expenditures totaling $7.1 *billions* of dollars. Not all of that was spent within the year, but it is all supposed to be out on contract by next June. Some contracts may extend for a considerably longer period of time.

These immense sums cover the following categories of activity:

Missile research, development and production
Construction of missile sites
Missile-related construction
Missile support equipment
Guided missile ships
Guided missile submarines
Missile detection systems
Anti-missile missiles
Expanded research and development efforts on military satellites and other outer space facilities

The various categories I have just listed suggest the obvious fact, which somehow escapes notice in many discussions of this subject, that most of the spending is directly related to military procurement and therefore to national defense. It summarizes the scope and type of effort considered necessary to defend our homes against aggressors. The fact that most of it relates to guided missiles only serves to point up the importance of these delivery systems in revolutionizing strategical and tactical concepts of land, air and sea warfare. In short, it is an example of the tremendous influence which technology is having in the military field, just as it has already influenced our economy and our daily lives.

Over and above the spending for strictly military purposes, an increasing amount of effort is being carried on in the space field. Here the distinction between what is entirely scientific and what is entirely military almost disappears. All the scientific disciplines are involved in the design and development of rocket weapon systems. The weapon systems themselves have become the carrier vehicles for our initial scientific space explorations. For example, the Air Force Thor intermediate range ballistic missile [IRBM], designed as a weapon, became the first or booster stage for the first lunar probe attempt. The Army programs with which I am identified have employed the Redstone ballistic missile, an operational weapon system deployed with NATO's Shield Forces, and the Jupiter IRBM, our version of the 1500-mile missile which the Air Force will operationally employ.

The modified Redstone became the first stage of our Jupiter C test configuration, which is the vehicle which launched all the Explorer Earth satellites and which, on its first flight in September 1956, achieved an altitude of 682 miles and a range of over 3,000 miles with an inert payload. We used the Jupiter as the first stage of our lunar probe shots and will use it for other space applications coming in due course.

These are but three examples of the close interrelationship between the military and scientific developments in space. I could mention others; in every case, the impetus comes from the military programs, just as it did in Europe. I would like to insert this remark parenthetically: it is not the scientist or the engineer who decides whether a rocket will launch a satellite or deliver an atomic warhead. If only we could achieve a lasting peace in this troubled world, the advances we accomplish as rocketry progresses would all have solely peaceful missions.

Unfortunately, we must consider human nature and politics as they exist—not what we might like them to be. So we must be prepared in advance to defend our way of life against any threat. Recent demonstrations of technological prowess by a potential aggressor have made the threat all the more real, and greatly reinforced his influence in political, military, economic and psychological, as well as scientific areas. If others intend to exploit space for aggressive purposes, we must have the capability of dealing with the challenge.

It is well to remember that space investigations did not begin with the launching of the Free World's first scientific Earth satellite by the Army team on January 31, 1958. Long before that, scientific studies of the higher atmosphere had been conducted by means of balloons and rockets. I took part in some of these tests during the post-war period at White Sands, New Mexico, at a time when people generally wanted to forget wars and wanted to enjoy some of the good things they had to forego during the conflict. Not too many people were interested in what we were doing, or in what could be done by way of space flight. Since then all three of the armed services, and many researchers and scientists in universities and research foundations, have joined the spreading effort. The Air Force and Navy, as well as the Army, have built up rocket development teams and facilities—in each case suited to the missions of that service. A vast amount of useful data has been accumulated.

In the engineering and production of rocket hardware, which is the presently accepted means of transportation from launching sites on to points in space, a great deal has happened—much of it in the last few years, and much of it based in principle and theory upon pioneer work in Germany. In retrospect, as future generations enjoy the benefits of this enterprise, much of what we have put together will seem pretty elementary indeed—just about as the first automobiles look to the fellow driving a 1959 model complete with air conditioning and powered gadgets of all kinds.

Looking back on the months since January 1958, it seems incredible that we were able to move as quickly as we did after the people and their government had decided to get into space in a big way. At the time of the decision, we had but one recognized space program funded for that purpose and in the hardware stage. This was the ill-fated Vanguard satellite project, a very ambitious undertaking from the technical standpoint, and one which, in my opinion, could have been a success had sufficient time been allowed for the thorough development and testing of rather advanced components and systems. There was other space hardware, but it was not developed expressly for the purpose. We had put together the Jupiter C to test our solution to the aerodynamic heating problem encountered when an object reenters the atmosphere from outer space. It saved us months in the Jupiter development and we achieved a solution, as the President reported to the American people when he exhibited a scale-model Jupiter nose cone on a nationally televised address in November 1957. That cone had been retrieved for us by the Navy after a Jupiter C launching three months earlier.

The other JUPITER C missiles were put on the shelf, so to speak, to await assignment. When we were given the go-ahead to launch a satellite, much of the hardware was ready.

Since the Free World's first satellite was in launched January 1958, we have seen continued progress. Among them have been the *Pioneer IV* space probe on March 3, 1959—the probe which passed the Moon and went on to become the first made-in-the-USA satellite of the Sun. And, of course, it is now a matter of history that one of our Jupiters carried the monkeys Able and Baker over a trajectory of some 1,965 space miles and were recovered in perfect condition.

One of the reasons for our space progress to date, I believe, has been the intelligent and aggressive direction supplied by the Advanced Research Projects Agency (ARPA) which was organized by the Department of Defense (DOD) early in 1958 to spearhead the program. ARPA has assigned our satellite and lunar probe missions, and similar undertakings to the Air Force, while the Navy was given the job of developing systems to track outer space vehicles.

The Congress and the Administration have established a new agency, called the National Aeronautics and Space Administration (NASA), which has now taken the reins on more of the scientific projects. NASA is the successor of the former NACA group, which made many notable contributions to aeronautics. While some of the programs have not been publicly disclosed, ARPA and NASA have announced certain planned projects in which we, the Army team, play a significant role. One of these is Saturn, the 1,500,000 pound thrust booster we are developing for ARPA. Another is NASA's Mercury, or man-in-space, project. Our contributions to this program will be to launch a number of manned capsules in the nose of Army Redstone missiles. These first manned flights are programmed to achieve an altitude of 80 miles.

Now let us examine some of the possibilities inherent in the further development of satellite and space technology. One intriguing possibility is that on the Moon, other planets, or the planetoids, we may find sources of scarce materials urgently needed in our expanding economy and technology. It has been estimated that an iron-nickel asteroid of normal meteoric composition having a diameter of only 200 feet would be worth one billion dollars at today's prices. No one is in a position at this juncture to predict specific-benefits—but we can be sure they will be there.

Another area is communications. It is possible, by the positioning of a satellite at an altitude of 22,300 miles, to have it remain stationary, in essence, relative to Earth's surface. For orbiters only a few hundred miles above Earth, like the Explorers, the period of revolution is in the order of an hour and a half to two hours. For a satellite at the distance of the Moon, the orbital period is 28 days. At an altitude of 22,300 miles the period will be 24 hours, which means a satellite revolves at the same speed at which Earth rotates on its axis. In effect, the satellite would then have a fixed position relative to the surface of the Earth underneath. Three such satellites, going around in the equatorial plane and being spaced 120 degrees apart as they circle through the same orbit, could be utilized as a permanent system of relay stations for high frequency radio and television communications. Since every point on Earth will always be in line-of-sight contact with one of these three satellites, and since each satellite can always see the two others, they would enable use to provide the whole surface of the Earth with an uninterrupted communications link for worldwide telephone, telegraph, radio broadcasting and television service.

Another potential involves weather forecasting. Annually storms cause hundreds of millions of dollars in crop damage, or in damage to communities, such as in Miami the other day, because we cannot adequately predict or in any sense control their occurrence. It is extremely difficult to determine from Earthbound vantage points the distribution of air currents and cloud structures over the entire surface of the globe in a very short period of time. This would inevitably lead to better forecasting and to better understanding of the causes of weather.

A third area is aid to navigation. With standard equipment one can now, by stellar observation, locate the position of a ship at sea within about one mile. By the same procedure, an aircraft can be located only within a possible error of five miles. When overcast renders the stars invisible these methods are useless. But with a satellite in orbit with a continuously operating radio beacon, or one which would transmit when queried by ship or plane, it would be possible to locate the position of either ship or aircraft within a few feet. I believe the practical economic values of such an aid are immediately apparent.

The scientist and researcher are motivated by curiosity, the desire for new knowledge, and the achievement of better understanding of the operation of natural laws and principles. Rarely, if ever, can he foretell the total impact of his discovery. For example, the curiosity of a few people in finding out what makes the Sun shine led to the demonstration of the thermonuclear reaction, and thus to the application of thermonuclear power—the most promising single discovery of our time.

I would like to believe, and to persuade you, that the equipment for peace and the incentive for peace may be found in the exploration of space, because in contemplation of the beauty of natural law illustrated there, all men may come to realize that they are in reality brothers, and that against the vastness of Creation they are really minuscule figures. As the cost of maintaining armaments increases, and as we appreciate the enormity of the effort we are now embarked upon to delve deeper into space, we may reach a point at which it is no longer economically feasible to disagree internationally, and as a consequence, understanding and peace my be forced upon the most recalcitrant nations. This indeed is the ultimate challenge presented to us by the Space Age.

*** END ***

— Chapter 16 —

LADIES AND GENTLEMEN OF THE MINNESOTA EDUCATION ASSOCIATION: [59]

I am a little uncertain as to how to approach this assignment. Judging by my personal mail, coming from all parts of the country, from boys and girls in elementary schools, high schools and colleges, our young people seem to know more about rockets and space vehicles than do the designers and developers of the Army's missile team!

They send me designs for ships to reach the Moon and other worlds. One boy even sent a sample of his solid propellant rocket fuel—and the material was so sensitive that it might easily have ignited anywhere along the route in Uncle Sam's mail system. By comparison, we fellows are still driving Model T Fords, while the youngsters are well on their way to building the solid gold Cadillac of the Space Age.

Sometimes I feel my daughters, after hearing their classmates talk, and seeing some of the latest fictional wonders in the movies or on television, must regard their old man as a back number!

So, with some trepidation, I will try to talk a little about space travel, asking you to remember that the problems and factors we must reckon with are a little more difficult to resolve outside the classroom.

As one who has espoused the cause of space travel for a long time, sometimes under rather trying circumstances—as when the Gestapo clapped me into a Nazi jail for talking about it—I would like to say that the enthusiastic reception it has received in America's schools has been most encouraging.

In spite of the seeming interest in the exploration of outer space, the program still requires salesmen. But the need is for salesmen who have some sense of responsibility, who know what they are talking about, and who are sincerely interested in the nation's best interest—not their immediate financial gain. This space business recalls a comment by Edward Carpenter which can stand repetition: "Every new movement or manifestation of human activity, when unfamiliar to people's minds, is sure to be misrepresented and misunderstood."

I suppose I have been asked a thousand times, "why should we explore space?" or "why do you want to go to the Moon?" Answering the second question first, I don't expect to be on the first space ship that reaches the Moon. If it turned out to be one of Mr. Khrushchev's vehicles, I am sure there would be no room for me. If it is one of ours, they will take a long look at me and say, "Pshaw—you are too old!" But I do expect, and I am looking forward, to getting there when travel between Earth and Moon has been stabilized and the ICC, or whatever passes for it in the Space Age, has established the usual tariffs!

As to the why of space travel, I would point out that the events of the last eighteen months or so have given us an even more valid reason. The Sputniks and the Luniks were convincing proof of the dynamic, large-scale effort which Soviet Russia is making to advance the development of rocketry for eventual space travel. They amply demonstrate that the Russians consider space travel as something more than a new area for scientific investigation. The Communists are keenly interested in the possibility of extending the already large sphere of political influence, and increasing their prestige and their military strength.

It is undeniably true that the Soviet achievements in satellite and space technology have heavily influenced world opinion concerning the Russian military, scientific and industrial potentials. Certainly they have profoundly affected the attitude of those smaller nations which must rely upon the major powers for military or economic support.

Consequently, I firmly believe that an active space program of major proportions has become a "must" for America and the western world, for we must check any further expansion of Russian influence in order to safeguard our way of life.

In the Age of Technology, into which science and engineering have transported us, it is extremely difficult to separate military and political considerations. The same ballistic missiles that launch satellites and scientific missions can be equipped with atomic warheads for purposes of aggression and conquest. Further, we must recognize the threat posed by the military exploitation of space ships and satellites by aggressors. If only for the requirements of national defense, we must support an all-out effort to achieve a maximum capability.

I do not wish to convey the impression that only political and military considerations enter into the concept of space travel. Expanding our knowledge of the spatial environment will enrich our understanding of the universe of which we are but a tiny part.

I can illustrate this by referring to the data obtained by some of our space exploration projects. The Explorer satellites launched by the Army in 1958 have returned valuable measurements of space phenomena. They provided exact information about the existence of a hitherto unknown radiation belt surrounding Earth. They confirmed the correctness of our computations regarding orbital data. They indicated the frequency of micrometeorite impacts. They furnished temperature recordings from which we deduced that it will be possible, without too much effort, to maintain a range within an orbital vehicle acceptable to human beings.

Our *Pioneer IV* space probe, likewise launched with the assistance of the Jet Propulsion Laboratory (JPL) of the National Aeronautic and Space Administration (NASA), became the first made-in-the-USA satellite of the Sun. With this firing, we demonstrated the capability to achieve escape velocity of more than 24,000 miles per hour. Radio contact was maintained with the probe over a range of more than 400,000 miles, using miniature transmitters with an input of a fraction of a watt, and enabled us to gather additional data of great value for future tests.

Of course there is much interest in the possibility of economic returns from the large investments required to carry out significant space flights. While the present state of the art is relatively immature, we can look forward confidently to a number of highly useful applications.

For example, an orbital vehicle can be utilized for continuous surveillance of Earth's total cloud cover. It could alert threatened areas to imminent storms. The data accumulated by this "all-seeing" weather eye could provide the basis for the most efficient weather forecasting service mankind has ever had available. If you think about the immense damage inflicted by storms annually, the loss of lives and property, any reduction in the toll would be very worthwhile. The entire world could benefit from a global weather forecasting and warning service that might well be undertaken as an international program.

We are currently engaged in developing the Saturn multi-stage space rocket system for the Advanced Research Projects Agency (ARPA) of the Defense Department (DOD). The first stage of this huge rocket will generate about 1,500,000 pounds of thrust, or five times more than our present ICBMs. With that kind of power it would be possible to place really big satellites, on the order of 15-20 tons, in practically permanent orbit. Suitably instrumented for communications purposes, a number of satellites could improve international transmissions markedly. Presently, transoceanic radio communication is often impeded and interrupted by atmospheric disturbances. Also, the existing undersea cables are subject to rupturing and their capacity is approaching the saturation point.

The transmission of television signals from one continent to another has been accomplished only with considerable difficulty. One of the most serious limitations is that high frequency waves proceed only in straight lines and will not bend effectively around the Earth.

It is possible, by positioning a satellite at an altitude of 22,300 miles, to have it remain stationary in essence relative to a point on the Earth's surface. For orbiters a few hundred miles above the Earth, such as the Explorers, the period of revolution is on the order of an hour and a half to two hours. For a satellite at that distance of the Moon the orbits period is 28 days. At the 22,300 miles altitude, the period will be 24 hours, which means the satellite would revolve at the same speed at which the Earth rotates on its axis. Three such satellites, going around in the equatorial plane and spaced 120 degrees apart as they circled through the same orbit, could provide a permanent system of relay stations for high frequency radio and television communication.

Since every point on Earth will always be in line-of-sight contact with one of the three orbiters, and since each satellite can always see the other two, they would enable us to provide the Earth with an uninterrupted communications link for worldwide telephone, telegraph, radio broadcasting and television service. The Saturn system can make this service possible.

Another potential economic application of space vehicles is an aid to navigation. Employing standard equipment, a navigator can, by stellar observation, locate the position of a ship at sea within about one mile. By the same procedure, an aircraft can be located within a possible error of five miles. If the stars become invisible because of overcast, these methods are useless. But with a satellite sending out a continuous radio signal, or which transmitted when queried by ship or plane, it would be possible to locate either ship or aircraft within a margin of a few feet.

The opportunity for unobstructed surveillance of the Earth from either satellites or space stations offers obvious military advantages, as well as benefits in the scientific and economic areas. It would help to improve the cartography of the Earth and to complete geodetic surveys. The exact location and dimensions of the continents could be defined with greater accuracy. Eventually we could register the continuous changes that are taking place on Earth's surface.

Space travel will, of course, offer unlimited opportunities for scientific research, the total results of which are beyond the bounds of our imaginations. The discovery of the Van Allen Radiation Belt was a most promising start. Future satellites, manned and unmanned, will further explore the mysteries of cosmic radiation, which may be the principle cause for mutations in the plant and animal kingdoms, although only the products of their collision with molecules of the atmosphere really impact the Earth.

We must learn more about the radiation emitted by the Sun. It is the mainspring of every form of life on this planet, but only a small portion of it passes through the filter of our atmosphere. Telescopes carried in artificial satellites will permit astronomic observations without atmospheric interference. Unmanned observation satellites will help to prepare for the launchings of space vehicles to the Moon and Mars.

Finally, we need to study the effects of the space environment on man himself, in order to prepare him for voyages into deeper space, for his coming visits to the Moon and the nearer planets. From these explorations we shall gain new knowledge about the history of our planetary system and the development of life—perhaps in its most primitive form.

NASA has embarked upon Project Mercury with the objective of placing man in orbit in a capsule. We are contributing to this program. Several Army Redstone missiles will be launched to test out the capsule separation and recovery systems. Finally, we shall give one of the astronauts a ride in a ballistic missile during which he will soar far beyond the atmosphere and travel across a part of the Atlantic Missile Range to descend into the ocean. From these experiments will come highly useful data preparatory to space travel.

Sometimes we hear complaints that there are many problems to be solved of vital interest to mankind and that they ought to be taken up before we busy ourselves with space exploration. I believe the United States is big enough to do both without harm to either. Whether some nations not yet engaged in space programs might wish to join in the effort is a question which may, to a certain degree, influence the pace of our progress. It will not, however, alter the natural course of space development.

Man employed ships to discover new continents. He used airplanes to conquer heights never previously attained. He will also take advantage of the possibilities inherent in rocket propulsion to use space vehicles to probe the unknown regions beyond Earth's atmosphere. Curiosity alone is sufficient motivation, even if much more is required to muster the necessary resources.

It should be apparent that in addition to designing and building space transportation systems, much of what we accomplish will depend upon the contributions of the medical profession. Pooling the knowledge and skills of the doctor and the engineer, we can create the artificial environment needed to ensure man's survival in the vacuum of outer space.

The deeper we penetrate into this new environment, the more we learn about it, the less often will anyone ask "why travel in space?" People will come to realize that extending our knowledge of the Universe is as natural and inevitable as the hazardous expeditions undertaken by explorers like de Gama and Columbus, or the tabulation of the natural elements. Judging from the rapidity with which space travel development has moved from the realm of science fiction into the disciplines of exact science and technology, I am confident that we shall move forward at a rate which may surpass our boldest expectations.

We cannot evaluate tomorrow's possibilities and requirements fully in the light of what we know today. But there is no doubt that the scientific exploration of unsolved phenomena through space flight will be a fascinating and challenging undertaking. It is man's greatest adventure and his biggest opportunity to contribute to world understanding and peace.

Let me conclude by stating one more commonly heard question—I might call it the $64 question, if you will permit another reference to the television industry. Will we overtake the Russians? I think we will, but must qualify that answer by asking another question—how and when?

I have given you my personal views on the Soviet intentions, which are no different, basically, than are their objectives in all other activities—to exploit space for Communism. This is not simply a race between competing teams of rocket designers. It is a deadly competition to achieve a position of dominance: we are motivated, as always, by a desire for peace and knowledge; they are motivated by a dream of empire as old as the Czars and Caesars.

I submit that the Russians have plainly evidenced what they are up to; they began an immense and well-organized rocket development program many years ago. They are, as it were, trying to leapfrog our advantage in aircraft by concentrating on missiles. They have power plants bigger than anything currently available to us, simply because they set their sights on space before we did. The American people must realize that you don't create something like the huge Saturn rocket overnight; it takes time, money, and a great deal of human experience and skill. Meanwhile, the Soviets are not standing still. They have built up an impressive momentum and they have no intention of allowing us to catch up and surpass them if they can beat us to the punch.

We are in a race for which the stakes are as big as space itself and I suggest we must spit on our hands, haul in our belts and get to work in earnest! The consequences of anything less are too awful to imagine—human freedom, human progress is the stake.

*** END ***

— Chapter 17 —

The first Germans arrived at Fort Bliss, Texas, in 1945, a little over one year after the last A-4/V-2 was launched in Europe. Between April 1946 and September 1952, sixty-seven A-4/V-2s were launched at White Sands proving Ground (WSPG). Two were launched at Florida and one from the aircraft carrier *Midway*. Von Braun and his team were transferred to Redstone Arsenal, Alabama, with the Army, in 1950. On February 1, 1956 the U.S. Army Ballistic Missile Agency (ABMA) was established at Redstone with Brigadier General John Medaris as commander. Its technical core consisted largely of the Guided Missile Development Division of Redstone Arsenal.[60] In 1960, the civilian agency, NASA, incorporated the scientific space program into its mission, located adjacent to Redstone Arsenal.

This speech was given to the Army Ballistic Missile Agency (ABMA) a few months before von Braun and his team transferred to NASA at Marshall Space Flight Center (MSFC). The MSFC was carved out of the Redstone Arsenal installation, making it contiguous to the Army's Redstone Arsenal where ABMA was located.

Tribute to General John Medaris and the ABMA Team [61]
30 January 1960

Some of you will recall my promise, on our third anniversary, that when the next ABMA (Army Ballistic Missile Agency) birthday rolled around, the assembly shop of our Fabrication Lab would be filled to the rafters with a rocket bigger than anything attempted before.

As the ensuing months went by, and reexaminations and the reevaluations, and the inevitable hemming and hawing came into the picture, there were times when I had serious misgivings about my role as a prophet.

But today those who toured the Fab Lab saw the first of our huge Saturn boosters. Hans Maus and his loyal people had to work overtime to do it—but they made it possible for me to keep my promise.

This is another proud achievement in the history of ABMA. It puts another star on the record of the great leader we honored today on the occasion of his retirement. I can say this General Medaris, no one gave you any bigger going away present than did the men and women of ABMA!

We can't let you take it with you. But in due time we are planning a little boat trip down the Tennessee River to carry it to Cape Canaveral—as close as we can put it to your vacation home! I suspect you'll be around to offer some helpful hints.

We who have been at your side during these last four years, when so much history was being made here, deeply regret our parting—though we know it was your wish and that you have fully earned retirement.

We know, better than anyone else, how much you have contributed to our achievements. Your courage, your fearless espousal of our programs, kept us together and kept things moving even when you didn't know where the money was coming from for the next pay day! I well remember the occasion when Government auditors came down to find out how you built satellite launching rockets when you had no mission to do so. I wonder how the official feels who ordered that check up—if he realizes that without those Jupiter Cs, there would have been no Explorer [satellite] to restore faith in America throughout the Free World.

You were the first officer in the entire defense establishment to plead the cause of the super booster—to warn the people that unless we developed a million pounds of thrust, we could never overtake our competition. At a time when it simply was not wise—so far as politics go—to talk about such things, you had the courage to voice your sincere convictions.

It now seems quite likely that before the next ABMA anniversary comes around, many of us will no longer be working for the Army. We earnestly hope that NASA will take a leaf from the book of our first boss. The kind of support we received from you, and from the entire Army through your influence, will be needed as long as we continue to move ahead in our space program. I would like to believe NASA will say, 'this is the way General Medaris would have done it, so this is the way to do it.'

When the rain came pouring down yesterday, I called General Medaris's staff and inquired whether the parade would be called off. The answer I got was, "Don't you worry, General Medaris has rejected three weather reports." I stood on the parade field this morning and heard the praise, respect and honors heaped upon you by Secretary Brucker, General Hinrichs and General Callahan. All of them and much more are due you—free men everywhere owe you a debt of gratitude for your imaginative and inspiring leadership.

Whatever you do in your new career—and I am sure it will be fruitful and rewarding—a part of you will always remain here in ABMA. You are one of us and we shall never forget you.

We have learned much under your wise tutelage—much of the art of management, the art of salesmanship, the art of inspiring people to deeds beyond their capabilities. You made us into a team; you gave us the facilities and equipment we needed; you gave us hope. It took a long while, but finally and inevitably you convinced the highest authorities in the government that this institution is a national resource. Maybe you were even a little too successful in making your point, for in the ensuing deliberations on how to put this national resource to the best use for the country, the Army lost it—maybe that was the price you paid for success.

We wish you good fortune as you leave Redstone—we earnestly pray that you and Mrs. Medaris will enjoy many years in good health.

We hope that whenever you feel impelled to do so, we may have the benefit of your counsel and inspiration. You may be removing a uniform but you will never remove your identification with the organization that brought you into being. On behalf of all the men and women of my division, I convey to you our sincere gratitude, our abiding affection, and our love for John Bruce Medaris, soldier, statesman and leader without peer.

*** END ***

The transfer of the space program from the Department of Defense to NASA took place in 1960[62]. The turnover ceremony was in front of Building 4488, the Army Headquarters at Redstone Arsenal, Alabama, July 1, 1960, at 9:30 a.m. NASA Director T. Keith Glennan said in a message to Dr. Wernher von Braun: "It is a pleasure to welcome you and your associates at George C. Marshall Space Flight Center to the NASA Family. We at NASA are proud to entrust to you one of the heaviest responsibilities of the national space program—the development of large, reliable launch vehicles to lift scientific experiments, and even man himself into space. The historic achievements of the von Braun team under direction of the U.S. Army have well equipped you to accomplish this exacting task. We look forward to our association with Marshall Space Flight Center. With a sense of dedication, we face the difficult task ahead, confident that your splendid record of accomplishment is just beginning."

Von Braun made the following remarks:

GENERAL SCHOMBURG AND GENERAL BARCLAY:

This is certainly a meaningful day in the lives of many of us.

I want you to know that for some time now we have approached today's event with some feeling of trepidation as well as anticipation.

Most of all, I think, we are swept by a feeling of gratitude and respect for the magnificent leadership that the United States Army has displayed for years in the conquest of outer space. Without the bountiful and courageous backing and support of the Army—and particularly the people of ABMA and AOMC—the Free World would not have jumped off into space nearly so soon. Nor would we have gleaned a little bit more of the knowledge of our universe nearly so promptly or effectively.

As we accept this transfer of the space portion of the Army, it really causes a lot of us to be swamped with memories of a whole bucketful of success—and a few set-backs, too—that we have shared together. *Explorer 1* for example. Do you remember that night General Barclay?

The Development Operations Division is proud too, of its contribution to this nation's missile progress.

And now that this country has seen fit to assign the Development Operations Division— officially and formally the George C. Marshall Space Flight Center ... all of us are looking forward to a long and fruitful association with the Army and contractor agencies here.

I think I can promise you that the Marshall Center will be one of the best neighbors you ever had.

We don't mean the kind of neighbor that just talks over the back fence once a week.

We're going to be a share-the-barbecue-pit, book lending, lawnmower swapping, let's-have-a-ball kind of neighbor.

As you know, we've got some work to do together, too. And I think you will find the Marshall Center just as helpful and friendly in that respect, also.

Like the Army, the Marshall Space Flight Center is going to continue to take an integral and active part in the community life of Huntsville and the Tennessee Valley.

General Schomburg, on behalf of Dr. Glennan and NASA Headquarters, it is with deep pride and with a sense of humility that I acknowledge and thank you for this official recognition today of the George C. Marshall Space Flight Center of the National Aeronautics and Space Administration.

And to my fellow associates and friends, I pledge to you my personal and wholehearted cooperation in creating, here at Huntsville, Redstone Arsenal and the Marshall Center, the finest and friendliest group of organizations and people that ever lived and worked together.

I know that all of us share the very deep conviction that the future holds a generous place for the great achievements yet to come from under this particular bit of sky just overhead.

But it will take everyone of us, working and living the best we know how, to get the job done.

Thank you.

After the transfer took place, von Braun pulled the Huntsville community together by creating an advisory committee. He wanted to be sure the residents understood what was going on at Marshall and make them a part of the process. As the newly developed powerful engines began to roar across the valley, the sound of which had never been heard anywhere else on Earth, he wanted them to experience some of the tests first hand by inviting them as witnesses. They would be able to express to the rest of the community how powerful the engines were and that the almost daily eruptions emanating from MSFC did not mean the world was coming to an end! The engines sent out vibrations that gave one pause for concern—windows rattled at great distances away from the stands. On the other hand, the tests also demonstrated that the program was on the road to success.

Another issue was the trepidation about growth. The population of Huntsville in the 1950s was around 17,000; it was to grow to well over 200,000 by the time *Apollo 11* was on its journey to the Moon. The business community was excited about the prospect for growth and the knowledge that Huntsville, Alabama, was the primary force in launching men into outer space. But, additional homes, schools, roads, and the infrastructure associated with that growth would have to be built at a very fast pace. He created a Civilian Advisory Committee. The first meeting was held July 11, 1960.

MR. CHAIRMAN AND MEMBERS OF THE COMMUNITY ADVISORY COMMITTEE OF THE George C. Marshall Space Flight Center:[63]

Well, as a good neighbor and a friend of mine said, "As long as neighbors meet and talk about each other while they're all together—instead of talking about each other some place else—they'll continue to be good neighbors."

Webster defines neighbor as: "A person who lives near another."

I'd like to go a little further and add that good neighbors live and work together in harmony and with understanding, each in his or her own way, working not only for the good of the family, but for the good of the community.

This is one reason Harry Rhett and I asked you particular members to serve on this committee: so we can sit down and talk together as friends and neighbors.

We know that each of you has contributed a great deal to the community over the past years, and we know that in your various fields of interest you have accomplished much for the good of Huntsville and Alabama.

We are very much aware that Madison County's newest entity, the Marshall Space Flight Center, is going to need a lot of local help from here on out. We don't need to mend fences yet; we're too young to have built any. But we are going to need a lot of advice, criticism and support—when we deserve it—from the members of this committee to make sure the Center earns a worthy place in our community, and retains that place. This is where we need your help most of all. We know how greatly we rely on the resources and people of Huntsville, and we sincerely trust that we can carry out our responsibilities to you in return.

Before going any further, I'd like to introduce to you a couple or so of my colleagues who are going to be working closely with you and me.

Dr. Eberhard Rees, my deputy director for Research and Development, and Mr. Delmar Morris, my deputy for administration. I might say that it's Mr. Rees's job to develop space vehicles, while it's Mr. Morris's job to find space around this place to build them in.

Also, here is Bart Slattery, my new information director, who came down from Washington today especially for this meeting.

Bart will join us officially on August 1, and he will be working closely with you. Whenever I'm out of pocket, call on Bart. I want you to know him well.

Harry Rhett has agreed that Bart serve as secretary for this committee. So if you want any work done, just ask Bart.

One of the fundamental purposes of forming this committee is to keep you informed. No one, I believe, has a deeper conviction than I have about how imperative it is that the community and the public as a whole know what we are doing, why we are doing it, and whether we are doing it badly or well. I hope, too, that by learning more about us that the public will understand us better.

Three members of our committee are from our local news media. They are Will Mickle, John Garrison, and John Higdon. They have played a big role for some time in keeping the public informed about the Center, as well as keeping you posted on a few developments around the world. We invite you gentlemen to spend more time with us and expand your coverage, if you think necessary, any time you like.

Having served on a few bureaucratic committees myself, I can say that we don't always come up with the desired results.

But in the case of our own Community Advisory Committee, I know all of us are convinced that we are going to see concrete and mutually-helpful results.

I think the main thing the committee members can do is to keep us on the right track in our dealings with the community. I'm very serious about this. When we get off base in working out common problems that will inevitably arise between the Center and the Community, we want you to let us know, and to advise us on a course of action. We anticipate that problems will arise from time to time in the area of traffic control, housing and schools. Even law enforcement, perhaps.

In some cases we may not recognize a problem when we see it. Again, don't let it slide. Advise us.

On certain occasions, when we might have some of our wheels down from Washington, we would like for you to meet with them. We would like to help put Huntsville's best face forward, and show off our good committee members.

One particular event that I'm personally pushing very hard right now may come off in September, and we will certainly need your advice and help to put it over. If we can talk the powers that be into approving this little shindig, we think it will be a big day for Huntsville and Madison County.

What I've proposed to Dr. Glennan is this: that we have a formal dedication ceremony for the George C. Marshall Space Flight Center. We would like to do it up big—invite General Marshall's old boss, President Eisenhower down, along with other assorted brass from NASA and the Defense Department. As part of the ceremony, we are proposing that Mrs. Marshall, General Marshall's widow, unveil a bust of the General. We don't know what luck we'll have, but we are trying very hard.

In addition to living and working with the people of this region, the Marshall Center is certainly going to continue its close and friendly companionship with the Army agencies at Redstone Arsenal. We are just as anxious to keep on being good neighbors with the Army and contractor groups as with everybody else.

I know that all of us in this room share the very deep conviction that the future holds a generous place for the great achievements yet to come from the people and institutions of this region.

But it's going to take every one of us, living and working the best we know how, to get the job done.

It goes without saying, that as far as the George C. Marshall Space Flight Center goes, we have taken a big step in whatever part we may play in this future by persuading you members of our Community Advisory Committee to join with us.

Thank you.

*** END ***

– Chapter 18 –

Doing Science

*All our science, measured against reality,
is primitive and childlike—and yet it is
the most precious thing we have*
— Albert Einstein (1879-1955)

Every aspect of Nature reveals a deep mystery and touches our sense of wonder and awe. Those afraid of the universe as it really is, those who pretend to nonexistent knowledge and envision a cosmos centered on human beings, will prefer the fleeting comforts of superstition. They avoid rather than confront the world. But those with the courage to explore the weave and structure of the cosmos, even where it differs profoundly from their wishes and prejudices, will penetrate its deepest mysteries.

There is no other species on Earth that does science. It is, so far, entirely a human invention, evolved by natural selection in the cerebral cortex, for one simple reason: it works. We must understand the cosmos as it is and not confuse how it is with how we wish it to be.[64]

Kepler and Newton represent a critical transition in human history, the discovery that fairly simple mathematical laws pervade all of nature; that the same rules apply on Earth as in the skies; and that there is a resonance between the way we think and the way the world works. Our modern global civilization, our view of the world and our present exploration of the universe are profoundly indebted to their insights.[65] In the following speech to the American Rocket Society, von Braun reaffirms Sagan and other scientists writings.

On Wednesday, June 7, 1961, a luncheon was held at the Biltmore Hotel in New York City, by the American Rocket Society, Inc., where Dr. von Braun was the guest speaker. Mayor Robert Wagner introduced von Braun. Dr. G. Edward Pendray, founder of the American Rocket Society, was the toastmaster. Distinguished guests on the dais were Mrs. Robert Goddard, Harry Guggenheim, General John Medaris, Dr. Richard Porter and Dr. Detlev Bronk.

Following are the remarks made by von Braun at the luncheon:[66]

SINCE OCTOBER 1957 WE HAVE FOUND OURSELVES IN A RACE FOR SPACE WITH THE Russians, whether we like it or not.

We are in a race for new insights, scientific insights in the mystery of the world surrounding our Earth and our atmosphere, and it is also a race for our military stature in this new environment. We should never forget that space, after all, is not a program but a place, and all these arguments about whether space has scientific significance only, or whether there is a future for the military in space, is rather meaningless in my opinion. It has never been possible to rule out altogether a space in which weapons can travel for military purposes. Right now, ICBMs, IRBMs and anti-missile missiles are traveling freely in outer space, so it would be rather ridiculous to say that space has no military significance.

On the other hand, we cannot possibly say that the United States today would be in mortal danger if we didn't occupy the Moon tomorrow. The Moon may become a militarily important objective ten or fifteen years from now, but today it most certainly is not. *Sputnik I*, which introduced us to this space race, came to most Americans as a shock, and there was a period of breast beating and questioning about what had gone wrong. We asked ourselves if our educational system was to blame for this; whether our scientists were inferior to Russian scientists; and even the moral fiber of our free society came into questioning. Belatedly, we finally embarked upon a space program of our own. We started out humbly enough. Our first satellite, *Explorer I*, weighed a mere eighteen pounds, compared to the several hundred pounds of its Russian counterpart. But in the meantime, our national space program—or rather its portion devoted for scientific explorations—has grown into a one-billion dollar a year program. I think this is an absolutely unprecedented phenomena; a national program, supported by taxpayers' money, which is not aimed at the production of consumer goods like automobiles and washing machines, or which is not aimed for any other commercial objectives; which has no immediate military significance; and yet, the taxpayer is expected to support it to the tune of a billion dollars a year.

Why do we have to support such a space program? I think the answer is that it is necessary to keep the machinery of our technological and scientific progress moving. I think in the last analysis what makes this machinery tick is very simple. *We have to continue to satisfy our curiosity. I think it is that simple.* Let me quote a few examples of how the desire on the part of a few people to satisfy their scientific curiosity has moved the world. Approximately twenty-five years ago, two Americans, astronomers or astrophysicists, I should say, became interested in the question of what prevented the Sun from burning itself out. They had calculated that under the terrific radiation energy losses the Sun was subjected to every day, every year, every decade, every century, the Sun should have cooled down to a dim, dark red in the many hundreds million years we know it existed; so they asked themselves: "what keeps the Sun hot? What mysterious mechanism is at work to furnish the Sun continuously with new energy?" They analyzed the solar spectrum and finally discovered that under certain conditions of temperature and pressure two hydrogen atoms could fuse into one helium atom, thereby releasing tremendous amounts of energy. Thus, thermonuclear fusion was discovered. Twenty-five years later, man succeeded in duplicating thermonuclear fusion on Earth. And when he did so, he succeeded in a very dramatic way. A little island atoll in the Pacific where the first hydrogen bomb was exploded vanished from the scene forever. Today, thousands of engineers and scientists are busy trying to harness this very same thermonuclear energy for power conversion. Their objective is to produce electricity from thermonuclear energy, and I am convinced that most of you in this room will still see the day when that dream will become reality. When we will be able to make the kilowatt hour for perhaps one-tenth to one-hundredth of what it costs today with abundant thermal nuclear power. Here you see the cycle closed—the cycle of scientific and technological progress. Two astrophysicists being curious about what keeps the Sun hot, the discovery leading to the most terrifying weapon ever invented. But in the wake of all this will be a bonanza for mankind—millions of electrical slaves for everybody.

Many, many other examples could be quoted. It was because an obscure Austrian monk by the name of Mendel became curious about the laws that govern heredity that the famous Mendelian Laws were discovered. Mendel experimented in his monastery garden in Austria with white and black peas, and he wanted to know the laws that govern propagation. The Mendelian Laws have now, a hundred years later, become the ground rule for all our efforts in growing sturdier and hardier cattle, and help us feed untold millions. It was because, sixty years ago, a physicist by the name of Roentgen became interested in the laws which governed the flow of electricity through as vacuum that x-rays were discovered. He had a little glass jar evacuated and ran his currents through the jar. Suddenly he discovered that in his blacked-out laboratory a plate in the corner of the laboratory room glowed every time he turned on the current. Some mysterious rays were apparently traveling between his glass jar and that plate. He placed all kinds of obstacles in the path of these mysterious rays and couldn't stop them. Finally he put his hand in the path and he saw the outline of his own bone structure portrayed against the plate. Thus one of the most powerful tools of medical science, both in diagnosis and therapy, was created. It was because a Britisher became interested in what happened to normal mold that penicillin was discovered, which in the meantime has saved millions of lives. What can we learn from all these examples? I think we can learn two things. One is that it pays off to satisfy your curiosity. The other is that it is impossible to predict what will follow in the wake of amazing discovery.

When Columbus took off on his immortal voyage, the purpose of the exercise was to improve trade relations with China. That problem has not been solved to this very day. But you will probably agree that Columbus did the right thing if you just look at the byproducts. Speaking of outer space, there are so many people today who ask this very same question. Why do we do all this? Why do we spend billions of dollars a year to satisfy our curiosity? Will we ever make the money? The honest answer is we do not know, but we can prove from past experience that it pays to satisfy our curiosity.

For the first time in human history, we now have the tools at our disposal to explore the world beyond the atmosphere. We have the powerful rockets to do it. We should put them to use. And that is my simple answer to the question a little old lady asked, "Why do we have to go to the Moon? Why don't people stay at home and watch television as the good Lord intended?"

I read with interest recently a column in the *New York Times* entitled "Space and Serendipity." Serendipity, I find, means the discovering of valuable or agreeable things not sought for. Serendipity, then, is one of the bonus results of all research, including space exploration. The value of research is always realized in the wake of discovery.

Another question that many people ask today is, "are we ahead of the Russians?" I think this was a good question two or three years ago. I think it is no longer a good question. Both the American and Russian space programs have developed into many different areas. It has become a very versatile field, and any sweeping generalizations to the effect that we are ahead, or they are ahead, have become rather meaningless. I think the true answer is, we are ahead in some fields and they are ahead in others. Let me be a little more specific. Let's take a look at the score.

I think we have fired more satellites and more space probes than the Russians. We have made the first successful attempts in radioing back to the Earth television pictures of the Earth's cloud structure in our TIROS project. This is a determined attempt to put satellites into use for better weather prediction. We were the first to demonstrate the feasibility of satellite repeaters, active and passive, for communication purposes over long distances. Our *Pioneer V* probe was the first to demonstrate the feasibility of interplanetary radio contact. Radio contact with *Pioneer V* ceased when this probe was twenty-two million miles away from Earth. We were the first to demonstrate successful recoveries from orbital flight with our Discoverer series, vehicles boosted by Thor/Able rockets. We were the first, I think, to demonstrate successful reentry tests with animals—our Able/Baker monkeys, which were fired 1600 miles down the Atlantic about two years ago. We at least published a demonstration to that effect. And finally, our rockets found the Van Allen Belt, this mysterious belt of trapped radiation which surrounds the Earth, and whose very existence was not even known three years ago, but which seems to be a deciding link [in] the action between solar radiation and many phenomenon such as radio propagation taking place on Earth.

On the other side of the ledger, the Russians have most certainly fired larger and heavier satellites and space probes than we. In their last attempt, *Sputnik VII*, which went into orbit a few days ago, was probably the most impressive demonstration along those lines. The Russians were the first to put any object into orbit with their *Sputnik I*. The Russians were the first to place animals in orbit and monitor their survival and behavior under their environmental conditions over an extended period of time. These animals were not retrieved, however. The Russians were the first to fire past the Moon into free planetary space with their *Lunik I*. They were the first to hit the surface of the Moon with their *Lunik II*. They were also the first to photograph the far side of the Moon, a trick that we have not succeeded in yet, although we have tried several times. We have all the indications that the Russians have also succeeded in recovering objects from orbits, and we know that they have a man-in-space program which is at least as aggressive as ours. And I can add to all this that the Russian space scientists that I have personally met in various international meetings impressed me as competent, dedicated and determined men. They are determined most certainly in the sense that they did not want to fall back in this decisive race with the United States for supremacy in outer space. The only area in which the Russians have a significant lead in the space program is in the area of large boosters. They have consistently lofted heavier payloads into orbit and into outer space than have we. Our shortcomings in this direction are my business as Director of the George C. Marshall Space Flight Center.

*** END ***

– Chapter 19 –

"Dr. von Braun, do you believe in God?" His response: "Yes, absolutely." Von Braun went on: "It seems to me your question is irrelevant. It is so obvious that we live in a world in which a fantastic amount of logic, of rational lawfulness, is at work. We are aware of a large number of laws of physics and chemistry and biology which, by their mutual interdependence, make nature work as if it were following a grandiose plan from its earliest beginnings to the farthest reaches of its future destiny. To me, it would be incomprehensible that there should be such a gigantic master plan without a master planner behind it. This master planner is He whom we call the Creator of the universe. One cannot be exposed to the law and order of the universe without concluding that there must be a divine intent behind it all… For me, there is no real contradiction between the world of science and the world of religion. While, through science, man tries to harness the forces of nature around him, through religion he tries to harness the forces of nature within him."[67]

IMMORTALITY [68]

TODAY, MORE THAN EVER BEFORE, OUR SURVIVAL—YOURS AND MINE AND OUR children's—depends on our adherence to ethical principles. Ethics alone will decide whether atomic energy will be an Earthly blessing or the source of mankind's utter destruction.

Where does the desire for ethical actions come from? What makes us want to be ethical? I believe there are two forces which move us. One is a belief in a Last Judgment, when every one of us has to account for what we did with God's great gift of life on the Earth. The other is belief in an immortal soul, a soul which will cherish the award or suffer the penalty decreed in a final Judgment.

Belief in God and in immortality thus gives us the moral strength and the ethical guidance we need for virtually every action in our daily lives.

In our modern world, many people seem to feel that science has somehow made such "religious ideas" untimely and old-fashioned.

But I think science has a real surprise for the skeptics. Science, for instance, tells us that nothing in nature, not even the tiniest particle, can disappear without a trace.

Think about that for a moment. Once you do, your thoughts about life will never be the same.

Science has found that nothing can disappear without a trace. Nature does not know extinction. All it knows is transformation!

Now, if God applies this fundamental principle to the most minute and insignificant parts of His universe, doesn't it make sense to assume that He applies it also to the masterpiece of His creation—the human soul? I think it does. And everything science has taught me—and continues to teach me—strengthens my belief in the continuity of our spiritual existence after death. Nothing disappears without a trace.

*** END ***

– Chapter 20 –

This speech was presented to the National Security Industrial Association at a conference focused on strengthening government and industrial partnerships.

I AM DELIGHTED TO HAVE THIS OPPORTUNITY OF MEETING WITH YOU TODAY AND I appreciate your thoughtfulness in inviting me. It is flattering to be asked to address such a distinguished group of industrial and military leaders.[69]

Since increased emphasis was placed on the exploration of space almost two years ago, and our role in the development of launch vehicles has expanded, it seems to me that I am spending more and more of my time wrestling with problems in management rather than technology. But don't be alarmed—I am not going to attempt to lecture such an experienced assembly of managers and executives on the principles of management.

Any discussion of management, as you all know so well, is apt to be dull, and the more you talk about techniques and principles, the duller it gets. Methods of management are no problem, anyway; the problems pop out when people get tangled up in the scheme of operations. And I feel certain, anyway, that you are that rare type of manager who can get a lot of work out of people and still keep them happy... or at least keep them from fighting in the halls.

Or you may be the type of executive who can pass out the money, assign functions, allot office space, authorize carpets, and reorganize the entire outfit without anyone losing face, quitting, or getting drunk on the job.

So, instead of management, I want to discuss the development of large launch vehicles for NASA's scientific exploration of space, and industry's role in that development program.

The space age is five and a half years old, if you date its birth from the Soviet launching of *Sputnik I* in October, 1957. Tremendous advances have been made in the United States and in Russia since that date, as one significant accomplishment has followed another in rapid fire order. Congress is confident—will soon grant NASA and the Department of Defense fiscal year 1964 appropriations for the exploration of space that will total more than seven billion dollars—we think. Space is here to stay.

Yet, a surprising number of people still ask: "What good is space?"

Queen Victoria once asked Michael Faraday what was the use of his experiments in electricity and magnetism. Experiments such as his are the basis of our electric power industry and of nearly everything else in our electronic world, although Faraday couldn't foresee all this at the time. But his confident reply to the queen was, "Why Madam, what is the use of a new born baby?"

He might have added: "It's a miracle of creation, a fascinating wonder, a link with the future in shaping man's destiny."

The political and economic spin-offs alone from a successful space program will more than justify its terrific cost. But the overriding reason for pushing as deeply into space as man can go is simply the satisfaction of scientific curiosity. Man's history is studded with discoveries that prove it pays in unexpected ways to learn all we can about ourselves and the universe. Aristotle said, "Man, by nature, desires to know." Man has an intellectual compulsion to ride over the next mountain to peer into the anatomy of the atom, or as President Kennedy has said, "To sail new oceans."

Space travel has long been an imaginative dream of mankind. Now, for the first time, our science, technology, and economy can support our imagination of the past. Because we can explore space, we must.

There will always be those who say, "it can't be done."

They are descendants of those who said Robert Fulton would never propel a ship by hitching a paddle wheel to a steam boiler. Or relatives of the people who said that Henry Ford was a fool for trying to put a gasoline engine on wheels, and that the Wright Brothers and Samuel Langley were crazy for putting wings on one.

Folks in our time have said that planes would never break the sound barrier. And that an object placed in space would never survive the "heat barrier" encountered on reentry into the universe.

Today they are saying that an astronaut can't pass through the radiation belts surrounding the Earth and live; or that his spacecraft will be punctured by meteoroids speeding through space. And even if he does learn to function in a weightless state for prolonged periods, a solar flare will get him.

Well, you know the answer. Man will surmount these obstacles to the exploration of space, just as he has overcome countless others during his progress from caveman to 20th century cosmopolite.

We have learned a lot about the Earth and our solar system during the last five years. The United States has launched a variety of Earth satellites and deep space probes, and has shared with the world the information gained about space and man's ability to function in this new environment.

During 1962, for the first time, the United States surpassed the Russian performance in total weight of satellites placed in orbit. At the beginning of this year we had orbited about 50 metric tons of payload, compared with about 41 tons for the Soviet Union, according to public announcements.

But we had to make four times as many launches as the Soviets to get our 50 tons of satellites into orbit.

During the first years of NASA's program for the peaceful and scientific exploration of space for the benefit of all mankind, we have somewhat ironically had to rely entirely upon rockets that were developed for defense. These have been modified for our missions and we have been highly pleased with the results. But, as you know, our ballistic missiles for defense do not match the weightlifting ability of the Soviet intercontinental ballistic missiles. They do not need to, because of our advances in weight reduction of atomic warheads. But, we have long needed larger, more powerful rockets for heavier Earth satellites, instrumented deep space probes, and manned space travel.

This need will be met by the Saturn family of launch vehicles now under development by NASA's George C. Marshall Space Flight Center at Huntsville. The civil service personnel at the Marshall Center have the know-how and the facilities for conducting a rocket program all the way from conception through design, development, fabrication, and testing. This unique capability has often been considered something of an anomaly, in government circles as well as in industry.

Questions raised about Marshall—and a few other NASA Centers—generally follow this pattern: Does the Center perform research and development work which would otherwise be contracted to private industry and universities? Does the government duplicate facilities available in industry, thus squandering the taxpayers' dollars? Does the government regulation stifle creative research?

A quick look at the activities of the Marshall Center and its in-house capability will demonstrate that such fears are footless. Our main mission at Marshall for the next few years is to provide launch vehicles to support Project Apollo, NASA's manned lunar landing program. In addition, we perform related research, and perform advanced system studies for space transportation concepts in the future.

While we are proud of our record of achievement in rocket development, and our team's present capabilities, we do not pretend to have all the answers. At the same time, we feel a weighty responsibility for the successful accomplishments of the missions assigned us by NASA. Space travel is challenging, complicated and costly. And it involves the lives of brave men. We have no choice but to use all the resourcefulness, inventiveness and ingenuity we can find. And we find a whale of a lot of it in industry. Here are a few "for instances."

During fiscal year 1962, about 90 percent of NASA's budget was spent by contracts as compared to 64 percent for fiscal year 1960. At the Marshall Center, 76 percent effort was contracted out the first years we joined NASA and in our proposed budget for 1964, only 7 percent will be retained in house for our own research and development operations, support, and administration, while a whopping 93 percent will be contracted.

We have had to expand our in-house effort at Marshall since the President said that we should go to the Moon—within this decade. Our laboratories are being enlarged, and new test facilities are under construction. But the launch vehicle and engine development projects assigned us have multiplied in number and mushroomed in size. When compared with our total effort, the portion performed in house has shrunk each year, while industry's role has steadily increased.

The three members of the Saturn family are Saturn I, Saturn IB and Saturn V. To illustrate their weightlifting capabilities, let me compare them with current launch vehicles. The most powerful rocket now in use is the Atlas-Agena B, which can lift 5,000 pounds into Earth orbit. John Glenn's Mercury spacecraft, which weighed about 3,000 pounds, was launched by a modified Atlas-D vehicle.

The Saturn I will be able to place the equivalent weight of seven John Glenn capsules into Earth orbit; the Saturn IB will lift the equal of eleven Glenn capsules, while Saturn V, largest launch vehicle now under development by NASA, will be able to lift a payload equal to 80 Mercury spacecraft.

Furthest along in development of the three is the Saturn I, which was launched for the first time October 27, 1961. Two other launches have been held since that time, each equally as successful as the first. The fourth Saturn I is on the launch pad at Cape Canaveral now, undergoing checkout for firing within a few weeks. Only the first stage is live during these first four research and development launches. Upper stages are inert, and about 100 tons of water is added to simulate full propellant loading.

The first stage of Saturn I is powered by a cluster of eight H-1 engines, burning liquid oxygen and kerosene. The Rocketdyne Division of North American Aviation, Inc., at Canoga Park, California, developed the H-1. The H-1 evolved from an earlier 150,000-pound-thrust Rocketdyne engine used successfully in the Jupiter, Thor, and Atlas missiles. The thrust of each engine will be uprated to 188,000 pounds, to provide a total of 1.5 million pounds of thrust for the booster.

The Saturn I booster is fabricated at the Marshall Space Flight Center. Marshall will build eight of these boosters for R&D flights. The Chrysler Corporation has been awarded contracts totaling more than 233 million dollars to produce, check out, and test twenty-one boosters. Chrysler's fabrication work will be performed in the 43-acre government-owned Michoud Plant in New Orleans. A Chrysler-built first stage for Saturn I will be the first completed hardware to roll from the Michoud Plant, sometime before the end of this year.

The Chrysler-built stages will be brought from the Michoud Plant to Huntsville by barge for static testing, and then to Cape Canaveral for operational launches, which are scheduled to begin in 1964. The Saturn I will test the command and service modules of the Apollo spacecraft in Earth orbit.

The second stage of the Saturn I is under development by the Douglas Aircraft Corporation at Sacramento, California. It will use six 15,000-pound-thrust RL-10 engines developed by Pratt & Whitney. These engines use the new fuel combination of liquid oxygen and liquid hydrogen, which gives us about 40 percent more thrust per pound of propellant than the liquid oxygen/kerosene combination.

Many severe mechanical and engineering problems had to be overcome in handling these cryogenics. As you will remember, liquid oxygen boils at 297° Fahrenheit, and liquid hydrogen at -423°. Douglas has successfully conducted numerous static tests of the SI-C engines in the second stage of Saturn I at the Sacramento Test Facility. Tests have been made for full flight duration [of] about seven minutes. And Pratt & Whitney Aircraft recently fired a single RL-10A-3 engine continuously for 28 minutes on an endurance test.

The second stage of the Saturn I will be flight tested for the first time in the fifth launch of this vehicle, scheduled for later this year. We are looking forward with a good deal of excitement, hope—and confidence—to this launch, for it will represent a significant milestone in the Saturn I program. It will be the first launch of the booster with the engine fully uprated to 188,000 pounds thrust each. Fins will be used on the booster for the first time for aerodynamic stability. It will be the first launch from our new launch complex 37, now nearing completion. And with the second stage live for the first time, the first Saturn I should put eight tons into orbit. Later launches will orbit more than ten tons. No payload this size has yet been injected into orbit from the Earth in a single launch.

The Saturn IB vehicle will be an interim step between the present Saturn I and the advanced Saturn V. The first stage of the Saturn I will be mated with the third stage of the Saturn V to give us a vehicle that can place about sixteen tons into orbit. With it, the entire manned Apollo spacecraft can be tested in Earth orbit. These flights will concentrate on crew training and perfection of module maneuvering and docking in orbit. On a lunar mission, the three modules of the Apollo spacecraft are stacked one on the other for launching. On the way to the Moon, however, the command and service modules separate from the lunar excursion module, and are reoriented with attitude controls, and turned 180 degrees to mate nose to nose with the lunar excursion module. This places the exhaust nozzle of the propulsion system of the service module forward. It can be fired in this position to brake the Apollo spacecraft to enter an orbit around the Moon. This separation and docking maneuver will be practiced in Earth orbit before a manned lunar landing is attempted.

The largest NASA launch vehicle now under development is the Saturn V. I have previously compared its size to the tallest building in town. But I understand that in New York there are 99 buildings that are taller, plus three in Brooklyn. So I have used the Statue of Liberty as a measuring stick. The Saturn V is our "Moon Rocket," the one that will launch the three-man Apollo spacecraft on a journey to the Moon before the end of this decade.

The first stage of the Saturn V is being developed jointly by the Marshall Center and the Boeing Company. The Marshall Center will build three boosters for ground tests, as well as the first flight booster. Boeing will produce 10 flight stages at Michoud. The design, development and production contract for this work with Boeing totaled 418 million dollars, and is the largest single contract ever awarded by NASA. This booster uses a cluster of five F-1 engines.

The F-1 engine is now well along in development by the Rocketdyne Division of North American Aviation. It burns three tons of the conventional fuel—liquid oxygen and kerosene—every second to produce 1.5 million pounds of thrust. This is equal to the thrust of all eight engines in the first stage of the Saturn I.

Rocketdyne has static fired the F-1 engine more than 250 times. In more than half a dozen of these recent tests the F-1 has demonstrated its full thrust of 1.5 million pounds for over the full 2½ minutes of operation.

The second stage of the Saturn V is being produced by North American Aviation, Inc., at Downey and Seal Beach, California. Units fabricated in California will be brought by water through the Panama Canal to a test site now under construction in southern Mississippi for static firing.

The second stage will cluster five J-2 engines, which will burn the liquid hydrogen/oxygen combination to produce a total of one million pounds of thrust. Rocketdyne has conducted more than 140 static firings of this engine. In a recent firing the J-2 produced its rated 200,000-pound thrust in a full duration static test.

A single Rocketdyne J-2 engine will be used in the Saturn V's third stage. Design studies of this stage are now being conducted by the Douglas Aircraft Corporation.

I hope that my description of the Saturn vehicles illustrates our reliance upon industry for major and important contributions to our development programs. Beneath each prime contractor, which I have mentioned, is a pyramid of subcontractors and vendors who support them in depth.

We are proud of our ability in rocketry at Marshall, but we have a near phobia against getting too large. We prefer to remain a moderately-sized, well-integrated organization. The Marshall Center designer can leave his drawing board and walk across the street to consult with the person who is doing component testing; or he can walk to the vehicle fabricator in the next block, while the test towers are just over the hill. We firmly intend to preserve and nurture a limited in-house capability. Otherwise, our ability to establish standards and to evaluate the proposals—and later the performance—of contractors would not be up to par. Selecting the right contractor is no easy matter. By keeping our hands dirty at the work bench, by keeping abreast of the problems and progress of technology, we feel we can merit, rather than command, the professional respect of contractor personnel.

Now, what comes after Saturn V?

NASA is engaged in a joint project with the Atomic Energy Commission to flight test nuclear rocket engines and demonstrate the practicality of nuclear rocket propulsion for space exploration. Marshall will furnish the vehicle stage known as the RIFT (Reactor In Flight Test) stage for flight testing the NERVA or Nuclear Engine for Rocket Vehicle Application. Marshall's major contractor for the RIFT stage is the Lockheed Missiles and Space Company of Sunnyvale, California. The RIFT stage will be flown as a third stage on the Saturn V in a typical trajectory. A nuclear rocket stage should increase payload capability by a factor of two or three for near-Earth missions, and even more for deep space missions.

Preliminary planning and studies are underway for Nova, a launch vehicle that will have several times the weightlifting capability of Saturn V. Nova would be used for advanced manned lunar and planetary exploration. Its exact makeup is not yet known. Coordinated system definition and preliminary design studies are being made by two firms, the Astronautics Division of General Dynamics Corporation and Martin Marietta Corporation.

Research in electric propulsion systems has advanced to the point that the United States will soon test fly four small engine models aboard Scout vehicles. While the thrust of an electric system is small in comparison with chemically propelled rockets, it has the advantage of sustaining its thrust for long periods of time with little fuel consumption. Electric space ships, once boosted from Earth with large chemical boosters, can be used for journeys to the Moon or other planets.

In addition to the accumulation of scientific knowledge, the exploration of space is yielding untold practical benefits. The fruits of a stimulated economy through the advances in technology are immeasurable. The discovery of new and better combinations of materials, processes and techniques of production required for space exploration have transformed industry, resulting in new and better consumer goods. Thousands of new jobs are created, exciting and worthwhile careers are made possible, and worldwide economic growth stimulated.

The feeble steps which man has taken into space so far represent only the beginning. No one can foretell exactly what the future will bring. But we may be certain of one thing—today's predictions will become tomorrow's accomplishments.

For as Jules Verne, who wrote of an imaginary trip to the Moon 100 years ago, said, "Anything one man can imagine, other men can make real."

*** END ***

– CHAPTER 21 –

Education in the post-Sputnik era was a constant theme throughout von Braun's speeches.

Our educational offerings must come under scrutiny since it is tomorrow's generation that will have to cope with the problems developing today.[70]

58TH ANNUAL CONVENTION ALABAMA LIBRARY ASSOCIATION
"CATCH AN EXPLODING STAR" [71]
APRIL 13, 1962

MAN'S ACCUMULATION OF KNOWLEDGE WILL DOUBLE BETWEEN 1960 AND 1967, according to a current estimate. This will be the fourth time our learning has doubled—and in an incredibly short time.

The yardstick for this measurement is the total knowledge man accrued from his cave dwelling days until 1750. This criterion had swelled slowly to twice its size by 1900; gaining impetus, it doubled the second time by mid-century; the third time came within a decade, as the frontiers of investigation were pushed forward in every field from 1950 to 1960

The amazing growth of knowledge during the sixties is a challenge to the professional librarian, who must catch this exploding star, and in the words of a popular song, "Put it in your pocket, never let it fade away."

Libraries have long been famous as the storehouses of knowledge, keeping us in touch with the world of the past. They are often called the memory of the human race, for on their shelves is the record of everything man has thought, dreamed or discovered. Their value to our civilization cannot be overrated when we observe that man, for all his inventive genius, cannot recall all the items on his wife's grocery list, from the house to the supermarket.

The demands of our time should dispel forever the antiquated image of the librarian as a nice but somewhat fussy old lady with no special training, who talks in whispers as she checks her books out and in; for our modern librarians play an increasingly important role in progress by obtaining reports of new discoveries and circulating them for immediate use. Your specialized knowledge in speedy classifying, cataloging, and circulation techniques and the use of advanced communication methods, reflects an emphasis on utilization rather than preservation. It demands a new breed of librarian, who is also a technical and scientific information specialist.

As knowledge increases—and hence the work of the librarian—your current shortage of adequately-trained professional personnel will become even more critical. I can sympathize with you. The National Aeronautics and Space Administration has just concluded a months-long nationwide recruiting drive for qualified engineers and scientists to help in the space program, and we still have vacancies. This fall NASA will begin support of a training program at ten universities to help meet our need for advanced graduates. Brain power is our most critical shortage in the national space effort.

I hope you saw some of our technical library facilities during your tour this afternoon. We at the Marshall Center are working with the Army to establish a joint scientific information center, which will be one of the largest in the Free World. It will be a principle source of library-type information on basic and applied research on the development and testing of rockets, their components and propellants.

This new center will be staffed by librarians, scientists and technical information specialists, who will analyze and evaluate technical information received from all over the world to determine its application to our missions. Knowing the requirements of their professional counterparts in the laboratories, they may often be able to find the answer to a perplexing research or development problem in current literature more rapidly than in the bottom of a test tube. Someone else may have already performed the same or a similar experiment. Two precious commodities are thereby saved: dollars and time.

Foreign scientific and technical publications will be screened and appropriate material translated and circulated on a far greater scale than at present.

We are also studying the feasibility of establishing teletype, facsimile, closed circuit transmission and other optimum communication methods between the main library and branches to be established at the elbow of our research and development worker. Automation techniques, including microfilm storage and retrieval, will be utilized more fully.

I am delighted that there are also better days ahead for our Huntsville Public Library, which is the oldest library in Alabama. If you librarians from all over the state think you have problems at home, console yourself by reflecting on the Huntsville library; like other Huntsville and Madison County governmental services, it has struggled valiantly to serve a population which has doubled more rapidly during the last ten years than man's accumulated knowledge. I share the hope of the library staff and friends of the library that it will soon move into planned new quarters, with improved facilities for expanding its valuable services in Huntsville and Madison County.

Your theme for National Library Week, "Read and Watch the World Grow," is well phrased advice. Everything we read depicts a world in ferment. Everything is exploding—knowledge, population, material goods—and man himself is bursting out into space. We are living through the greatest revolution humanity has ever known—the revolution of scientific change. Nothing like it has ever happened to us before. Not even the wisest pundits and commentators even agree on a name for our dawning new era. Some say it's the age of automation, others the age of medical miracles, the atomic age, or the age of space exploration.

I am reminded of the fable of the six blind men of the Orient who went to an elephant for the first time. The man who grabbed its tail declared that the elephant is like a rope. The one who felt its trunk argued that it is more like a snake; while the man who ran headlong into its side swore that the elephant is like a wall, and so forth. You remember the story I'm sure. The observation of each man was limited, and so was his conclusion.

All our observers of the current scene agree, however, that ours is an age of promise as well as problems, and that the years ahead can be the most fascinating, the most challenging, and the most abundant that the world has ever known.

From my viewpoint, I like the title "The Age of Exploration." I want to speak to you briefly on this exciting subject.

The space age burst upon a startled world with the Soviet launching of *Sputnik I* on October 4, 1957. Until that fall, proposals to send a manned expedition to even our nearest neighbor in space, the Moon, were regarded by most people in the United States as harmless fantasy or sheer lunacy. Today these scoffers are getting used to the idea that a lunar voyage, long a dream of mankind, will soon be an actuality. In less than five years the United States has successfully launched 71 satellites and deep space probes, and the Soviets 18. Unmanned space launchings have grown commonplace and men have orbited the Earth. The skeptics admit that the impossible has once again become possible. Though reluctantly admitting its possibility, there are still some people who consider the manned exploration space as impractical, prohibitively expensive, or downright unnecessary.

Every one of us is affected, directly or indirectly, by the effort to investigate, explore, and make use of outer space. It is the biggest scientific and engineering endeavor that the United States has ever undertaken. As such, its activities affect our economy, science, technology, education, and welfare. It is a national program that must have the understanding and support of all our people—not only through their dollars, but also through a vigorous personal interest in what we are doing, and how well we are doing it.

More and more of our citizens show this interest. They only ask: "How soon can man go to the Moon?" and "How will he get there?" Let me try to answer these questions.

First, the time is not far off. In May of last year the President proposed to Congress an accelerated space program which included sending a man to the Moon by 1970. We'll go sooner if we can. But first we have some immense problems to solve, much planning, working, and learning to do before we blast off.

Planning a trip to the Moon is far tougher than planning an automobile tour of the far west, and having to take with us everything we will need until we get back home again—including all of our gasoline, oil, food, water, and even the air we breathe. Our family car wouldn't haul all of our supplies and equipment. And that's one of our biggest bottlenecks for a manned lunar landing and return to Earth. Our current launch vehicles won't lift the necessarily huge payload past the clutch of Earth's gravity. There are other problems too, but none impossible of solution.

Before looking at some of them, let's review the purpose for such a trip. Why would man want to go to an airless, pock-marked, desolate orb 238,000 miles from Earth? It's not a pleasure trip, nor is it planned to achieve a spectacular space first, televised worldwide. This is primarily a scientific search for new knowledge. It will be an extremely strenuous and perilous journey through the strange and hostile environment of space. Because the Moon has no atmosphere—no wind or water to cause erosion—its surface has changed little in billions of years. The history of its creation has been preserved on its surface for a much longer period than man can find on the growing, changing face of the Earth. Not only the surface, but also the internal structure of the Moon may provide clues to the early history of the solar system and to the birth of the planets. There we may find clues to the origin of the universe—and life itself.

Although no man has been there—so far as we know—we have formed some definite opinions about the Moon through centuries of long distance observation. These beliefs need to be confirmed, and new knowledge of its surface and environment gained through close-up observation. Before manned lunar flights are made, we must use scientific instruments to blaze the trail man will travel. There may be hazards in space whose existence we do not yet suspect.

Our first close-up observations of the Moon's surface will be made with unmanned payloads sent there on one-way flights to report back their findings. Sometime this year a Ranger vehicle will surely hit the Moon with a little over 100 pounds of instruments. We hope it will relay information all the way, including close-range television pictures of the Moon's surface.

The first three of nine Ranger probes have already been made, but unfortunately we have not hit the Moon. The next Moon probe vehicle is at the Cape now, being readied for launch within a few days.

Sometime next year, in the Surveyor project, NASA will try to soft land 350 pounds of instruments on the Moon. This initial Moon laboratory will drill into the lunar surface and analyze the rock or dust samples it obtains. It will then automatically relay its chemicals and physical analyses back to Earth.

In Project Prospector NASA will land a vehicle which can wander about the surface of the Moon like old-time gold prospectors in the west. Ranger, Surveyor and Prospector series of probes should resolve many theories about the Moon and give us much unpredictable new information to aid us in manned lunar landings.

NASA's program for the manned exploration of the Moon, known as Project Apollo, is directed at the Washington level by D. Brainerd Holmes. The spacecraft being designed to carry three men on a lunar trip is much more elaborate than the one-man Mercury capsules in which Alan Shepard, Gus Grissom and John Glenn rode on their historic voyages. In a few years we will look back on the Mercury spacecraft as an historic but crude spacemobile—like a raft, a sled or a Wright Brothers airplane.

The Apollo spacecraft will weigh 75 tons, and be as tall as a five story building. It will have a stage for lunar braking and landing, a return stage, and the command module which houses the crew.

To lift this huge spacecraft into direct flight from the surface of the Earth to the surface of the Moon would require a giant launch vehicle. The powerful Nova, whose first stage will generate more than 12 million pounds of thrust, will be designed for this purpose.

From six months to two years may be trimmed from our departure date if the exacting techniques can be mastered for rendezvousing our spacecraft with a rocket stage in a low-Earth orbit. The two would be joined while orbiting the Earth at 18,000 miles per hour, and the rocket stage would then hurl the spacecraft on a trajectory toward the Moon

For the rendezvous method, two advanced Saturn launch vehicles, now under development by the Marshall Center, would be used. The cluster of engines in the first stage would burn liquid oxygen and kerosene. The second stage would use a new and powerful fuel combination—liquid hydrogen and liquid oxygen. The frigid hydrogen/oxygen mixture will give us 40 percent more thrust than conventional fuels.

A hydrogen/oxygen engine will be flight tested for the first time at Cape Canaveral in the near future. The two engines in the second stage of the Atlas-Centaur vehicle will generate 15,000 pounds of thrust each. Much larger hydrogen/oxygen engines are under development and will be used in the advanced Saturn and Nova vehicles.

Getting to the Moon first will be a spectacular achievement for the nation that wins the race. The political and psychological benefits will be tremendous. But the possible scientific rewards will be more durable. Getting there first is only one part of the contest. Two other factors are just as crucial. First, what will we learn from our explorations beyond Earth? Second, how will we use this new knowledge.

We want to be first, and are running a good course, although we started this particular phase of the race with a five year handicap on the size of our launch vehicles. But we are keenly aware that the glory of getting there first will fade quickly unless scientific information of a high order is obtained in the process, and put to good use. After all, Columbus discovered America—but we speak English.

We are counting on you librarians to assemble the resulting new knowledge and make it readily available to all who are interested.

*** END ***

– Chapter 22 –

Huntsville Ministerial Association
St. Thomas Episcopal Church, November 13, 1962 [72]

Hardly a day goes by that we scientists do not see long-held theories about the Earth, the Moon, and the stars shattered or confirmed as a result of space exploration. It's been happening ever since the space age began. *Explorer I*, the first U.S. satellite, confirmed a belief that an immense field of radiation surrounds the Earth. It plays an important role in the transfer of the Sun's life-sustaining energies to the Earth. *Explorer I* confirmed its existence and recorded more information on the high intensity of this radiation, produced by high-speed particles and their impact upon matter with which they collide.

Other satellites have run into large concentrations of protons rushing out from the Sun during solar flares. *Explorer X*'s special instruments also found low-energy protons in constant quantities, suggesting that the Sun emits them all the time, and not just in times of severe storms on the Sun's surface.

We have found that space is not a vacuum, but is filled with many kinds of rays and particles.

For centuries men have been fascinated by the phenomenon known popularly in our hemisphere as Northern Lights. We now regard the Aurora Borealis as but one manifestation of the intricate electron and proton particle phenomena in the Earth's outer atmosphere.

Even *Vanguard I*, the tiny 3¼-pound grapefruit-sized satellite, made far-reaching discoveries. It showed that the Earth is slightly pear-shaped, and that the equator, far from being a perfect circle, is wavy. These findings indicate that the structure of the Earth must have more strength and rigidity than once thought, to maintain its shape. *Vanguard I* is changing our ideas on the makeup of both the surface and the interior of the Earth.

And *Explorer XI* made suspect the steady state theory on the origin of the universe, once regarded highly by many scientists.

Today's technology gives us remarkable mechanisms for reaching deep into space. Astronomers have reported recently that young hot stars yield far less ultraviolet radiation than once believed. As a result they have drastically revised their estimates of how hot these stars really are, how fast they burn nuclear fuel, and how much ionized hydrogen there is in the galaxy.

I could cite many other important discoveries resulting from probes we have made thus far into the area beyond our sensible atmosphere.

Each new bit of information impresses us anew with the perfect orderliness, the beauty, and the tremendous immensity of the universe.

Our Sun is only a lesser light in 100 billion stars in our galaxy. And our galaxy is only one of billions of galaxies populating the universe. That makes our Sun one of a hundred thousand million billion Suns—a number that staggers the imagination.

Before such majestic splendor we stand in humble, reverential awe. The insignificance of our own solar system, puny in comparison, is obvious.

It is difficult to believe that the Creator of such a universe would confine all sensible organisms to this comparatively tiny planet.

The more we are permitted to learn of the wonders of this universe, the more firmly do we become convinced of our oneness with spirit. And the more we appreciate the kinship of science and religion.

Science and religion are the most powerful forces shaping our civilization today. Through science man strives to learn more of the mysteries of creation. Through religion he seeks to know the Creator.

Neither operates independently. It is difficult for me to understand a scientist who does not acknowledge the presence of a superior rationality behind the existence of the universe, as it is to comprehend a theologian who would deny the advances of science.

We do not expect to find, through the exploration of space, tangible proof of the existence of God. But as scientists we cannot but admire His handiwork more deeply as we learn more about creation. And indirectly we learn more about the Creator.

Man can never hope to comprehend an all-powerful and infinite God. Any effort to visualize Him, to reduce Him to our limited comprehension, to describe Him in our language, beggars his greatness. The search to know God better must continue in our time, as it has since man's creation. I find it best through faith to seek God as an intelligent will, perfect in goodness, revealing Himself in the world of experience more fully down through the ages, as our capacity for understanding grows. There is no reason to believe that the revelation process has atrophied. It responds to man's desire and capability for increased understanding.

For spiritual comfort I find assurance in the concept of the Fatherhood of God. For ethical guidance I rely on the corollary concept of the brotherhood of man.

We in the United States today find ourselves engaged in a fierce struggle for technical supremacy with a nation whose government openly flaunts its independence from the spiritual. Progress is claimed by the Soviets on a purely materialistic basis, with sole reliance upon man's strength and ingenuity. Such spiritual poverty is pathetic.

The science and technology of today are wonderful, but there are far greater and more durable considerations in man's eternal search for excellence. Science has unleashed powerful energies, and has revealed amazing possibilities for man's advancement. And space as we think of it today, as the next step away from the Earth on which we have lived so long, is not the last frontier. Over the horizon there will be frontier after frontier as man's God-given curiosity prods him to continue his investigations into all things material and spiritual.

Where will his path lead?

Will it be destruction or continued understanding and advancement?

Man has been given the power of decision between good and evil, and his choice certainly determines his future. It is fortunate that he does not have to rely upon his own strength and wisdom. Science and religion are sisters: Science—Harnesses the forces of nature around us; Religion: Harnesses the forces of nature within us.

We in America are blessed with a deep spiritual heritage. In our excitement of discovering new truths, we must not discard a previously-held concept until it has been proven unequivocally erroneous.

We can only speculate on the effects of space research on religious thought. It should certainly strengthen our concept of God as Creator, and the brotherhood of man, when considered wisely. Closer ties will undoubtedly be established among God-fearing men.

A basic similarity of belief exists in many religious faiths. Similarities will increase as spiritual truths unfold. But let us not forget that religion is an individual matter, not a spiritual commune. Man does not fit a common mold, either physically or spiritually. His soul is unique. Any attempt must always start with the individual.

We are rapidly approaching a period of religious crisis in the world. Religious authority will be questioned, doubts arise, and faith shaken in many congregations.

You must lead your people to deeper spiritual insight into man's purpose on Earth, into a clearer concept of the Creator, a closer fellowship with his brothers, and an unwavering faith.

As molders of religious fervor, you must shake off all doubt, hesitancy, or despair. Speak out boldly, with power and enthusiasm. Reason and logic are essential for firm, steadfast convictions, but don't let them rout all emotions from religion. Don't be ashamed to stir the hearts of worshippers. Religion is concerned with the tenderest and strongest emotions man can feel.

A man would be a robot without a heart that can feel joy, love, grief, compassion, or devotion.

We scientists are helpless to say whether our discoveries will be used for good or evil. And that is the tremendous decision that is to be made in the world today. We have mastered enough of the forces of nature to usher in a golden age for all mankind, if this power is used for good—or to destroy us, if evil triumphs.

Help America to rediscover its spiritual heritage, to relive it, reinterpret it, and believe in it. Your success will undeniably influence the course of future events.

The ethical guidelines of religion are the bonds that can hold our civilization together. Without them, man can never attain that cherished goal of lasting peace with himself, his God, and his fellow man.

*** END ***

– Chapter 23 –

"The Making of a Missile" [73]

NEW AND UNUSUAL EVENTS HAVE ALWAYS EXCITED OUR CURIOSITY AND CAPTURED our imagination. Our first day at school, first airplane flight—and first love—are experiences which impress us deeply, and remain clear in our memories. With repetition and the passage of time, however, a similar occurrence loses the freshness which originally thrilled us, and becomes a more routine, though no less significant, part of our daily experience.

Missile firings once attracted the attention which men bestow on first events. When the Army Ballistic Missile Agency launched America's first satellite with a modified Redstone missile in January, 1958, there was something of a sensation throughout the Free World. So many satellites have been placed into orbit during the 44 months since that date that the space around the Earth is crisscrossed by spent rocket motors and silent but still orbiting payloads; most of us who are not directly connected with the space program have lost count of them.

An ordinary missile firing, with no space mission, is so routine now that it is overshadowed in news media by the threats to world peace in Berlin, Katanga, the United Nations, and Laos. Such recurrent threats of aggression spawned the huge missile program of our Armed Forces, however; and I feel certain that you have no small amount of curiosity as to how a missile comes into being to counter them. For this reason, I would like to describe briefly for you the various steps necessary in "The Making of a Missile."

Before I begin, I must explain that the missile itself is only one part of a complete modern weapon system. As such, it is far removed from the family musket which the American of Revolutionary Days carried off to war, with a powder horn and a sack of rifle balls slung over his shoulder. A missile weapon system is the most effective method which our advanced technological society can devise for bringing to bear the destructive power of the weapon against a possible aggressor, anywhere on the globe. The missile is only part of a weapon system which includes numerous items of ground support equipment and skilled personnel. The people, the hardware, and the techniques for its use are combined into a self-sufficient unit of striking power for strengthening our Armed Forces and those of the Free World.

The yeast that leavens the bread of missile making is the human factor. Brought into being for the sole purpose of supporting the soldier in battle, a missile is the product of the combined ideas, skills, and diligence of thousands of craftsmen and professional people in laboratories, factories, and proving grounds across the nation. There is no typical missile man among them. In Massachusetts a woman technician working in an air-conditioned, dust-free, and isolated booth and clad in gloves and gown that make her look more like a surgeon than a factory worker, bends over a microscope to assemble a microscopic gyro. In Texas another smock-clad workman sits behind a reinforced concrete wall and looks through a periscope arrangement of mirrors while a molasses-like propellant mixture flows into an upright rocket motor casing. In Utah an engineer works late into the night to redesign a missile launcher assembly for better performance. Each one is making a vital contribution toward development of a missile system. The measure of success in the program is the sum total of the opportunities given each of the thousands of persons involved to contribute his best directly toward completion of the end product.

A missile concept may originate in either of two areas: by a military person who recognizes a threat which must be countered or the need for a specific weapon; or from a scientist or engineer who recognizes the possibility of a technical breakthrough which must be exploited.

An Army commander might say, "We need a modular missile system which can be carried on a standard Army vehicle, with no section weighing more than 10,000 pounds so that it would be air transportable. It should be easily and rapidly assembled, checked out, and fired by a small crew, but capable of delivering nuclear and conventional warheads from 25 to 100 miles with pinpoint accuracy."

It is the staff responsibility of the Combat Developments Section of Continental Army Command to prepare a formal statement of these military requirements for submission to Department of the Army. The Army publishes the approved requirements in a volume called the Combat Development Objectives Guide, where they serve as guideposts to U.S. research and development firms in industry.

This is the point at which the military man and the scientist meet for discussion of the concept—the Army to present its need, and the scientist to discuss the possibilities of achieving it under the present or predictable state of the art of rocketry.

Now let us go back to the origin of the idea from a scientist or engineer outside the government and bring his proposal forward to this same point. The scientist who sees an application in missilery for a technical breakthrough goes to the Department of Defense in search for a sponsor and funds to pursue his idea further. The Defense Department, always receptive to ideas for improved weapon systems, consults with its three branches of service to determine the need for the proposed system. If no need exists, the idea is of little value at that time for defense purposes.

Proposals referred to the Army go to one of its seven technical services, through which Army research and development is accomplished. An idea for a missile system is considered by the Ordnance Corps, which is responsible for development of the Army's arms, ammunition, and armored vehicles. The Army Ordnance Missile Command at Redstone Arsenal is the Ordnance Corps' field installation which exercises weapon system management responsibility for the nineteen missiles and rockets currently in the Army's arsenal. The Army Ballistic Missile Agency under the Command's direction, has commodity management responsibility for ten of these systems. For the purpose of simplifying the discussion, let us assume that the proposal from industry was for a ballistic missile which is our responsibility at ABMA.

Feasibility studies are conducted to determine the goals possible of attainment. When the decision is reached that the results will meet the expressed need, a time-phased development plan is prepared for taking the concept from the idea stage through to a completed tactical system. The development cycle includes the functions of research and development, production, training of troops, and preparations for field support of the system. Checkpoints are established along the development calendar, with definite dates selected for testing of components, for test firing of the first R&D prototype, for the initiation of production, for opening the first training classes. The time when all actions are to be completed, and the system is to be ready for deployment by troops skilled in its operation and maintenance is known as the Ordnance Readiness Date.

A successful plan depends on the realistic prognostication of development. While you cannot arbitrarily schedule a technological breakthrough, it is possible to predict a continuing advance in almost any field, based on past accomplishments, if efforts, funds, and emphasis are continued on the same scale. To be realistic, the scheduling must be limited generally in scope to the development area of the research and development spectrum. The plan should begin with design studies, component and prototype development, rather than basic research. The developer is not asked to take an untested theory of aerodynamics, propulsion, or guidance and incorporate it into an operating subsystem or component. Research on such principles is done prior to inclusion in a system.

With the development plan completed, the next step is to survey the available resources and facilities, and determine the allocation of work among them. The basic decision lies between how much work can and should be done in the government's own facilities—called in-house—and how much should be contracted to industry. Several government agencies and installations in addition to Redstone Arsenal have facilities and personnel uniquely qualified to contribute to the making of a missile.

The Atomic Energy Commission, of course, supplies nuclear warhead components. The Ordnance Special Weapons Ammunition Command contributes to the development of warhead sections. The Army Signal Corps is the source for battlefield communication equipment. The Ordnance Tank–Automotive Command at Detroit supplies standard Army trucks and trailers and engineering support for specially-designed vehicles to give the system cross-country mobility. The Army Corps of Engineers furnishes proven components such as air conditioners, heaters, gas turbine generators, and fire extinguishers.

The Army's own facilities are limited, however, comprising, in fact, less than 10 percent of the total effort in our missile programs. We depend upon industry to perform the bulk of our work for us, under contract. Many private industrial firms have developed specialized skills and knowledge which enable them to make significant contributions toward the making of a missile. Thiokol Chemical Corporation, for example, has a long and outstanding record in the development of solid propellant rocket motors. Rocketdyne has specialized in liquid-fueled engines, while other firms are proficient in developing guidance systems, radar, test equipment, and numerous other components.

ABMA has employed different methods for securing the help of contractors in its missile programs in the past. It has retained weapon system management responsibility and technical supervision of the program, and served as its own prime contractor to coordinate the contributions of other government agencies and industry. In this approach we have relied heavily on our in-house research and development capability and managerial ability.

In other cases we have retained only the weapon system management responsibility and technical supervision of the program, selecting a prime contractor for the entire system. He, in turn, has enlisted the support of a wide variety of subcontractors and vendors, who furnish the prime contractor with entire sections, assemblies, or subassemblies. The firms which supply these items respond to the prime contractor, whom we hold responsible for the overall design, development, and performance reliability. In this approach we are buying management services, as well as hardware.

Within the established policies of the government, the factors that determine which arrangement will be used are the urgency, the comparative costs, the location and caliber of the technical knowledge, the existence and need for sustaining in-house competence, and current in-house workloads.

Let us assume that the decision is made to select a prime contractor for development of a new missile system to meet our stated requirements. A broad invitation is issued to industrial firms to attend a bidder's conference. These invitations are issued through the headquarters of eleven Ordnance districts into which the United States has been divided. The headquarters for one of these districts is located here in Birmingham.

Interested companies send their representatives to ABMA, where they are given detailed briefings on the military requirements. Discussions are held, opinions and questions flow back and forth. The industrial representatives then return to their companies, and prepare their proposals. Then all the proposals are returned to ABMA, where they are studied and evaluated. The one which fills the need in the most effective and economical manner is accepted, and contract negotiations are begun. The awarding of a contract is influenced by the ability of the firm to initiate immediate research and development, preferably in existing facilities under his control, and subsequently to enter into production. The prime contractor must be able to perform a substantial share of the work himself, and should be able to exercise firm managerial and technical control over a number of subcontractors and vendors. Since the prime contractor for a huge development program is normally a major manufacturer, it is important to encourage maximum subcontracting, to distribute procurement dollars throughout the national economy, to maintain the industrial base, and to stimulate competition. This is accomplished through an aggressive small business program, through contractual terms to this end, and other measures.

Once the prime contractor has been selected, one of the first tasks is to establish supervisory relationships between our agency and his company, between his company and the first-tier subcontractors, and between our agency and other participating government installations, tying all together in a close-knit working relationship. The leadership, or impetus, must be given [to] the project by our agency. This requires all the management techniques of organization, integration, coordination, communication, and control. Yet we must give the prime contractor side latitude to exercise his own management function, within the boundaries of existing government policies and procurement regulations. As weapon system manager, we retain the responsibility for major decisions, with approval and veto power over important actions of the contractor, and are responsible for enforcing statutory restrictions on government procurement. While we feel that we must monitor the contractor's operations sufficiently to evaluate his performance, and assist him wherever desired, we try not to bog him down with unnecessary reports and regulatory procedures.

Now, with the requirements for the system documented, development plan adopted, the people who are to perform the task selected, and a management structure established, we are ready to produce some hardware. There are a number of characteristics which we wish our finished product to possess, in keeping with the tradition of the Armed Forces for supplying our troops with the best possible weapons and equipment. Chief among them are: reliability, mobility, simplicity, ruggedness, accuracy, versatility, maintainability, and safety.

Reliability must be designed into a system from the outset. If there is a reservoir of components and assemblies which have been tested until their performance and reliability are known, they can be drawn upon for application to the new system. The use of these proven components obviously contributes to the overall system reliability, and shortens development time, as well. But reliability is not synonymous with conservatism in design. It may come from radical departures from standard approaches. Perhaps the greatest single factor affecting reliability is the basic quality of the engineering insight, ingenuity, and judgment exercised by competent individuals on the development team in the choice of approaches, components, and their assembly into a workable system. Confidence in the reliability of a new weapon system is established only through carefully planned and instrumented testing of components, assemblies, and complete systems.

Mobility and ruggedness are kindred characteristics demanded of all weapons and equipment on a nuclear battlefield. The Army has changed its organization, weapons, doctrines, and tactics to assure its survival and continued action under nuclear attack. The emphasis is on more mobility, greater flexibility, and better communications. Our Army must be able to move, shoot, scoot, and communicate—and this mobility must be built into its missile weapon systems, from the smallest to the largest.

The Army may be called upon to counter aggression anywhere in the world. It must be able to move quickly, on a minute's notice, and go into combat immediately. This requirement implies air transportability and cross-country mobility on the ground, for highways and bridges cannot be relied upon to remain intact for movement of heavy weapons and equipment.

The missile must be rugged enough to withstand the shock and vibration of movement and rough handling to which it is subjected on the ground, and during launch. It must also withstand the environmental conditions of heat, cold, ice, dampness, dust, and dirt, without malfunctioning.

Versatility is characterized by the ability of the missile to carry either a nuclear or non-nuclear warhead, and to place it on target at widely varying ranges.

Concern for safety of the crew members centers around the handling and arming of warheads and the high-energy fuels. The trend toward premixed and preloaded fuels is directed toward eliminating the hazardous handling of liquid fuels at the launch site. It also promises more efficiency at environmental temperature extremes.

Refinements of basic concepts should always strive toward simplicity, the elimination of gadgetry. The benefits are ease of manufacture, ease of operation, ease of maintenance—and lower costs. We have underway at ABMA and in the plants of our contractors an active Value Analysis-Engineering program, directed toward obtaining the greatest value for each dollar spent. Value Analysis techniques are applied throughout the development cycle in an effort to simplify hardware, methods, and techniques, while maintaining the same or improved performance.

Simplicity in operation and maintenance would also tend to minimize the amount of training required to ready a unit for deployment with the system. Human factors engineering is relied upon throughout design and development to keep operation and maintenance procedures well within the limitations and capabilities of man's mental and physical abilities, to provide a wide margin of safety for men using the system under the stress of battle.

These then, are some of the desired characteristics we seek. It is obviously impossible to attain the optimum in each of them in a single missile system, for one might be the antithesis of others. For instance, greater accuracy might be obtained by adding more refined guidance and control systems. But to do so might sacrifice the objectives of simplicity, ruggedness, and ease of operation and maintenance. Within any one missile system, therefore, there must be trade-offs between opposing characteristics. The user, who is our customer, strongly prefers a well-rounded, rugged, reliable system with acceptable accuracy to an extremely accurate but overly-sophisticated and cumbersome one which is operable only under controlled conditions by laboratory technicians.

The task of producing a complete system which embodies these characteristics is broken down into two main categories, the missile and its ground support equipment, each of which is further divided.

The three main parts of the missile are the propulsion system, guidance package, and warhead section. The ground support equipment usually consists of a launcher, firing station, communication pack, power pack, and test and maintenance equipment.

Warheads are divided into nuclear and non-nuclear, or conventional, classifications. Progress [has been made] in reducing the huge size of the first atomic warheads [such] that even our smaller rockets, such as the Little John and Davy Crockett, are nuclear tipped. If the missile or rocket can also carry conventional explosives or a chemical warhead, it presents the battlefield commander with a versatile weapon for annihilation of any target within his assigned area.

The guidance and control section of the missile, commonly called its brains, is crammed with electronic and mechanical instrumentation. Here the industrial craftsman meets exacting standards which test his skill. The precision demanded in guidance and control components is exacting, for a slight flaw might cause the payload to miss its intended target by hundreds of yards.

When the term "lubricant" is used, most of us think of an oily, viscous fluid motor oil, for example, but very likely not air. Yet the demands for frictionless bearings for ultra-precision gyroscopes, where the slightest drag to free rotation may cause errors in the missile guidance, has led to the development of bearings in which air serves as the lubricant. It is easy to understand how air, with its extremely low viscosity, works wonderfully as a lubricant. It is also easy to understand how the slightest surface imperfection, or bit of foreign material, may cause actual contact of the surfaces separated by the thin film of air, with resulting high friction. There is one reason for the extremely high precision, measured in millionths of an inch, required in these components [consider the following]: sometime when you're a passenger in a car starting up from a stoplight, close your eyes and try to estimate the speed built up and the distance traveled as you move off, using no other information than the pressure exerted against you by the back of the seat, and the passage of time. This is the only information the missile's inertial guidance system may use to compute velocities and distance traveled, yet it must measure several hundred times more accurately than your car's speedometer and odometer, which have the advantage, for measurement, of wheels rolling in physical contact with the road.

Motors are developed and thoroughly tested by static firings before inclusion in the missile. The solid-propellant motor is loaded at the factory and arrives in the field storable over an extended period of time, or ready for instant use. The mixing of the ingredients and pouring them into the motor casing, except for the danger involved, resembles the mixing of a devil's food cake. It goes into the oven and cures into a rubbery or plastic-looking compound.

A fire control set of a missile system, such as that for the Sergeant, feeds guidance instructions and other data into the missile, adjusting it for accomplishing its mission. The firing set checks out the missile to determine its launching readiness, and performs an automatic countdown for firing it. The Sergeant missile's firing set includes a computer which calculates the guidance settings when it is given the location of the launcher and the coordinates of the target.

During research and development the missile and its ground support equipment are thoroughly proven in a planned series of flight, operational, and environmental tests.

The first series of flight tests is made under scientifically controlled conditions, to verify the feasibility of the concept, and to determine if the new system possesses all the intended design characteristics. Redesign is expected to follow these early test rounds, when necessary. In an effort to integrate all components of the missile and ground support equipment into a weapon system, firing tests include the propulsion system, the guidance and control package, the air frame, the launcher, and as many other of the units of ground support equipment as possible.

The entire system can be tested only by firings, but this is also the most expensive method of testing. For this reason, the utmost care is taken in planning and executing the launchings, to obtain the maximum amount of useful data from a minimum of firings.

When the engineering series of tests is completed, a series of service tests is conducted for the user of the system. When these tests are completed satisfactorily, the system is ready to be procured by modern mass-production methods.

In most cases, the contractor awarded the production contract is the same one who has performed the research and development effort. His selection is almost imperative for a smooth transition into production and to save time in the development schedule. Not only does he already have in his possession most of the drawings and some of the facilities necessary for production, but he also has a group of people who know the system. And there is no clear-cut dividing line between the two functions, but instead, considerable overlapping. Industrial engineering for making the system more susceptible to production techniques, the acquisition and equipping of facilities for production, tooling up, and even the beginning of production, may be accomplished before research and development is entirely completed.

This overlapping of the research and development, procurement, and training functions, commonly called telescoping, is absolutely essential for shortening the development cycle and thereby reducing "lead time"—that time interval between the expression of the original concept and the issue of new equipment to troops. With such a complex weapon system as a missile, the lead time required may vary from two to ten years, a rather long time to the user who originally expressed the need for such a system. If time can be saved by overlapping, or telescoping, the functions of documenting a requirement, conducting feasibility studies, selecting a developer, building prototype hardware, testing and redesigning, procurement, training troops, and preparing for field support of the system, it behooves us to exploit this technique to the fullest.

Telescoping can shave from one to three years from the development cycle, thus combating obsolescence and providing the most advanced weapon to the user within a time frame when it may be expected to combat the predicted threat for which it was conceived.

Although time is gained, the price of telescoping is sometimes high. The decisions to go into early development and then into early production are made with [the] calculated risk that elaborate and expensive changes will not be required.

The production of missile weapon systems has caused a trend away from mass production and toward the provision of fewer, more complex weapon systems. The primary cause of this complexity is the dependence on electronic devices, which themselves grow increasingly intricate and complicated. Many of these devices are new products, based on a new and expanding technology, and require production facilities with special equipment not in existence for commercial products.

In cases where the contractor does not have available all the facilities necessary to produce an entirely new production item in sufficient quantities, the government has had to finance the construction of new facilities to insure the success of its program. Because of the unique nature of their contribution to a system, many subcontractors also require additional facilities.

Sufficient additional facilities are constructed or acquired to provide a capability of producing the system at the desired rate. The schedule of production is based on maximum use of facilities and available skilled personnel, and is also geared to the need for arming units for deployment.

One of our primary concerns during production is to assure that the quality of the finished product is maintained at desired standards. The contractual document requires the contractor to have an acceptable quality control system. His program is reviewed by our people to determine its adequacy.

Because we are pushing the state of the art in order to combat obsolescence, thereby working more than ever with marginal designs, we must increase our quality control effort proportionately. Several tools are needed to conduct a successful quality control program. First, the production contractor should complete the design of the test equipment he is to use and develop his quality control capability while the system is still in the research and development phase, before production begins. The test equipment should be so designed that it can be used in his plant, or elsewhere, in the event another firm should later be awarded a production contract. The same equipment can be used in the field support phase, after production is ended.

For the same reason, that breakout to another contractor may occur, the prime contractor is required to reduce his test procedures to writing, in an approved format.

The quality control manager should be so located in the organization that he participates in decisions weighing scheduled production against quality, and has the authority to stop production if quality is not up to the required level. The quality, reliability, and performance of production models are continually verified during procurement by flight testing rounds set aside for this purpose.

My discussion thus far has dealt with the efforts to design, build, test, and produce the hardware of a missile system. But you will remember that I said it is not a complete weapon system until it is deployed by trained troops. Without them, the hardware which has been so painstakingly developed would be no more than an expensive assortment of complicated junk littering the battlefield.

The complexity of the system makes the early initiation of training imperative. For the electronic portions of the equipment, for instance, a basic course in electronics at the Army Signal Corps' school in New Jersey may be a prerequisite, followed by more specific instruction at the Ordnance Guided Missile School at Redstone Arsenal. The Ordnance Corps is responsible for training individuals in maintenance and repair, while the Artillery trains in operation and use of the system. After the troops acquire their Military Occupational Specialty, or individual job training, they are brought together for unit training.

On a new system, the contractor usually conducts classes on New Equipment Training for key personnel, some of whom become instructors for troop training courses. The principle of telescoping is applied here, also, and R&D versions of the missile are allocated for training purposes, so that troops may be familiar with the system when the first tactical missile comes off production lines.

The payoff for our development efforts comes when the missile weapon system is deployed—but our responsibility and our efforts do not end at this point. As weapon systems manager, our final function after making a missile is to keep it made—that is to support that system as long as it is in the hands of troops anywhere in the world.

We must make certain that the logistic system funnels serviceable materiel to troops on a timely and sustained basis. This task is complicated because the individual parts in any one system may number several thousand, by the geographic spread of deployed units, and the different users. The cataloguing of parts and publication of supply documentation is a tremendous undertaking in itself. The logistics plan must provide for efficient management of the receipt, storage, and prompt issue of all materiel needed to assure maximum service life of the missile system.

Although the development cycle is near its end at this point, it has turned a complete circle, for there is an integral feedback of performance data from the field to research and development for application to the next and still more advanced missile weapon system.

From this description, I hope I have left the impression with you that the making of a missile is a tremendous and diversified undertaking. To meet the needs of the military, and to obtain the greatest advantage from scientific breakthroughs, and the soundest application of technological advances requires the most complete coordination of ideas, skills, facilities and resources. As a weapon system manager, we have a heavy responsibility to provide the highest degree of objectivity and impetus to our development programs. To do so requires the highest management skills, ability, and continuing experience. Our managerial talent must include not only scientific and engineering personnel, but also people experienced in all phases of planning, procurement, and production. I believe that it would be safe to say that the Army Ballistic Missile Agency, by its past achievements, has demonstrated its competency to administer a coherent missile development program. It is our desire to continue to discharge faithfully our present assignments, and those of the future.

*** END ***

– CHAPTER 24 –

Alabama A & M University of Huntsville, was organized in 1875. It is an historically traditional land-grant school.

THE MEANING OF SPACE EXPLORATION [74]
A & M COLLEGE, HUNTSVILLE, ALABAMA

PRESIDENT MORRISON, DEAN EDMONDS, AND DR. BELCHER, I AM DELIGHTED TO BE here this afternoon to address the students, faculty of the A&M College, and members of the Kappa Alpha Psi Fraternity.

The one thing about which we can be sure as the Space Age unfolds is that man will benefit greatly from this vast new medium that he has entered. Precisely what we will gain can be defined no more completely now than could be envisioned at a similar early stage the dividends from the early steam engine, the discovery of oil, the Wright Brothers demonstration that men could fly through the atmosphere, or the ultimate benefits accruing from the release of nuclear energy.

All these things have altered, and are altering, the patterns of civilization on our planet. The same applied to the broad and accelerated program we have undertaken for exploring space near the Earth and out to vast distances from the Earth, and for assessing and measuring the forces, particles, and chemistry of nature.

The subsidiary benefits of the national space program will be increasingly felt in educational institutions as well as our technical-industrial complex. More and more, thoughtful people are regarding space for what it truly is: a rich resource—just as land, rivers, seas, the atmosphere, and knowledge, skill, and brain power are great resources. The exploration of the limitless resource of space is an objective so tremendous and meaningful that nothing short of effort on a national scale could do it justice.

You know, almost everywhere I go, people are still asking me:

Why do you want to explore space?
Why are you guys so set on going to the Moon?

The ultimate goal of space exploration is pure scientific inquiry, the satisfaction of man's innate curiosity to learn all he can about the nature of things—of himself and the universe.

For centuries imaginative men have dreamed of a trip to the Moon. But for the first time man's science, technology, and economy are able to support the imagination of the past for manned space travel. Because man *can* go into space, he must.

I believe you understand this. Whenever I talk with college students and teachers, I feel that I have a sympathetic audience. I believe it is because you live largely in the world of ideas. Many of you are dreamers. And probably more of you than will admit have your eyes on the stars.

Mainly, though, you are curious. You wonder about the great and mysterious universe around you. You want to know what makes it tick, where it began, what it's made of, and where it ends. And what is the ultimate destiny of man, a puny creature, who has performed miracles by using his brain power to control powerful forces of nature?

A trip to the Moon may provide us with answers to some of these questions. Because the Moon has no atmosphere—no wind or water to create erosion—its surface has changed little in billions of years. The history of its creation has been preserved on its surface for a much longer period than we can find in the growing changing face of the Earth. Not only the surface, but also the internal structure of the Moon may provide clues to the early history of the solar system and to the birth of the planets.

The Moon has also been impacted by particles from outer space, which would have burned up in the Earth's atmosphere. Examination of these particles may help us to unravel other mysteries of the universe.

The Earth's atmosphere is a blanket over us, shielding us from deadly radiation and particles from space, and providing an environment that permits life to flourish. But it is an obstacle to astronomers who want to study the stars and the heavens. Our atmosphere, thin as it is, either absorbs, reflects, or distorts most of the frequencies available for probing of distant stars and planets. If the astronomer's instruments were based on the Moon, he would get a view of the majesty of the heavens as if a pair of dirty glasses had been removed from his eyes.

We know that there are numerous hazards to manned space flight. These include radiation in the Van Allen belts, danger from solar flares, and speeding meteoroids. And we expect that other hazards will be revealed in space and on the Moon and planets as we continue our inquiries with unmanned probes. But these must not deter us. Fear of the unknown has been one of man's greatest weaknesses during his long and difficult struggle upward.

The exploration of space will require huge sums of money. The NASA budget which the President has recommended to Congress totals more than five billion dollars. This is a lot of money. Some people are apprehensive that the emphasis on space will be detrimental to other pressing needs of man and society. The United States has the resources and capacity to support a strong space program, and at the same time carry on its undertakings at a reasonable level. The cost of the NASA budget to each person in the United States is less than 75 cents a week. This is less than the amount we spend on cigars and cigarettes.

And as Mr. James Webb, NASA Administrator, has said more than once, "Each dollar spent on space research will put two dollars back in our pockets."

The emphasis on space will not siphon away support of research in other fields. In fact, the tempo is increasing in numerous fields from the stimulus of the space program. Many experiments are being conducted in the fields of physics, geophysics, astronomy, and biology.

Although pure scientific inquiry is sufficient justification for our space program, there are many good, hard practical reasons for the program as well. Successes in space exploration bolster the international prestige of the United States.

The broadest practical payoff comes from the improved technology. Advances in scientific knowledge spur advances in technology, and those in turn stimulate economic development. New materials, new products, and improved manufacturing processes are developed. New and better jobs are created, and better living follows.

The dollars devoted to our space program are not launched up to the Moon, to vanish forever. They are spent right here on Earth—in the nation's factories, workshops, and laboratories. They pay salaries, buy equipment, and supplies. And the money passes like a round robin from person to person, vitalizing the economy.

Probably the greatest indirect benefit of space research is its stimulus to education. Better trained and better educated workers are demanded in the Space Age. A fresh look has been taken at our entire educational system since *Sputnik I*. The hard core subjects of English, mathematics ands science are given more stress in our elementary and high schools. At the college level it called for the training of more engineers and scientists, an improved curriculum, and better qualified teachers.

For years we have cited increasing need for engineers for our technological society. But we have failed to lick the problem. Figures recently released by the engineers' joint council show a drop in freshman enrollment in engineering colleges, and a decline in bachelors' degrees granted.

In the United States, our colleges graduated 52,700 engineers in 1950. This number dropped to 37,800 by 1960, and is expected to decline to 32,000 by 1965. While output of engineering graduates continues to decline, the Soviet Union continues to graduate more than 120,000 engineers annually. These contrasting figures have serious implications for the future.

NASA has a broad program, planned in depth over the years ahead. It will need the support of large numbers of adequately trained personnel. We must renew our appeals to high school students to choose engineering as a career, so that this downward trend may be reversed without delay.

Among the intangible benefits of space exploration are its effects on our social and cultural life. Barriers of time, distance, language, race, creed, and color are crumbling as enlightenment spreads. The concept of the brotherhood of man is gaining more universal acceptance.

The discovery of new scientific truth has meaning only as it effects man and his society. None can deny the scientific progress of the Space Age is profoundly affecting man's mental and spiritual outlook. It may stretch man's thinking to unprecedented heights, and cause him to reexamine some of his basic values and attitudes.

As a result of space research, man can become a finer creature in a more perfect society.

Now, where do we stand in the exploration of space today?

We've just crossed the threshold of man's greatest adventure. Our astronauts have not traveled 200 miles from the surface of the Earth. They have compared their orbital flights to a ride in an extremely high altitude airplane. The Earth was always close by if needed for an emergency landing. But soon our astronauts will be launched deeper into space. They will watch the Earth receding, growing smaller and smaller, until it resembles a tiny ball in the distance. Then our traveler will realize that he is entirely enveloped by space!

Our first tottering steps across the threshold of space can be likened to the beginning of air travel in 1903. When the bicycle wheels of Orville Wright's 12-horespower plane touched down in a field at Kitty Hawk, North Carolina, he had flown for 59 seconds at a top speed of 30 miles per hour. No one then foresaw from this feeble beginning the jet travel of today, the huge aircraft industry, and the affects of air travel on our way of life.

Since the President said that we should explore the Moon and the planets, we have made much progress in packing our bags for the trip to the Moon, first stop on our extraterrestrial travels.

Lunar Orbital Rendezvous has been selected by NASA as the most desirable method for the first lunar attempt, from the standpoint of time, cost, and mission accomplishments.

The Advanced Saturn has been chosen as the launch vehicle for starting the Apollo spacecraft on its lunar trip. Development of the Saturn family of vehicles is the responsibility of the Marshall Space Flight Center here in Huntsville. The Advance Saturn is the largest rocket currently under development in the United States. Its first stage will cluster five huge F-1 Rocketdyne engines, each one of which is as powerful as all eight engines in our first Saturn. With the addition of more powerful upper stages, the Advanced Saturn will have ten times the weightlifting capacity of the first version.

Expansion of our laboratories and test facilities at the Marshall Center is now underway for work on this powerful rocket. We have also selected the Michoud Plant in New Orleans for the manufacture of booster stages for both Saturn C-1[75] and the Advanced Saturn. Static test facilities will be constructed nearby in Southwest Mississippi.

Three Saturn C-1s have been launched from Cape Canaveral in our research and development program. So far all firings have been just about perfect. The fourth Saturn left Huntsville for Cape Canaveral two weeks ago for its scheduled launch in the spring.

The manned Spacecraft Center has been moved from Virginia to Houston, Texas, for its expanded role in manned space travel. Houston will provide the Apollo spacecraft and will train the astronauts.

We are often asked, "Do you ever plan to use women astronauts in your space program?" As my friend, John Glenn, first American to orbit the Earth said, "The astronauts are all for it."

Much planning and preparation have been done for our first trip to the Moon. Months and even years of continued hard work are ahead of us before we transform this once-thought-impossible dream to reality.

Every American has a role in making man's long-cherished dream of space travel come true. And every one will share in the feedback of curiosity which prompts the journey.

Would your grandfather—or even your father—have believed as a youth that in his lifetime man would fly an airplane at a speed of 4,000 miles an hour? that nuclear-powered ships and submarines would cruise the oceans for months without refueling? that he could sit in his living room and watch a football game in California, and with a flick of his wrist turn and watch a United Nations debate in New York? that man would circle the Earth in ninety minutes? or that he would soon land on the Moon?

These are the technical advances, and you will see many more of them as the dawn of the Space Age brightens into a full day. Remember, their influence spreads, sometimes slowly, sometimes violently, into many other areas. Following in their wake may come problems of adjusting to new ways of life. Human beings often resist change, clinging tenaciously to entrenched habits—especially adults. Sometimes we seem confused, frustrated, and even angered by an advance that upsets our normal routine or outlook.

Some of our adults have been dragged screaming into the Space Age.

Students have less difficulty in accepting the fact that only change is constant. I believe that teenagers, more than any other age group, recognize the swiftness and the depth of the revolutionary change which the world is now undergoing. I find young people are by no means dismayed, but on the contrary, they are stimulated by the changes resulting from the expansion of our knowledge on every frontier.

Who can foretell what the future will hold?

One thing is certain: today's predictions will become tomorrow's accomplishments.

As Jules Verne, who wrote a fictional story of a trip to the Moon 100 years ago said, "Anything one man can imagine, other men can make real."

I hope you students, and even faculty members, will make a practice of believing in—and doing—impossible things.

*** END ***

— CHAPTER 25 —

BOYS SCOUTS OF AMERICA [76]
RAYMOND JONES ARMORY, HUNTSVILLE, ALABAMA

I AM EXTREMELY PROUD TO BECOME AN HONORARY MEMBER OF ONE OF THE FINEST organizations for boys in the world. Thank you very much, Mr. Mitchell, for your presentation.

The Boy Scouts of America can justly be proud of a long and successful history. Over the years it has helped to instill a foundation of ideals and beliefs on which ambitious boys can build worthwhile careers. From your ranks will come the men who will direct this country and carry out its aims.

Some of you will surely become scientists, engineers, and, perhaps, astronauts of the near future. As you probably know, twelve of our first sixteen astronauts are former Boy Scouts.

By the end of this decade, it is quite possible that former Scouts—brave men, who as boys had *your* training, wore *your* uniform, and won *your* awards—will explore the surface of the Moon. This will be a difficult and hazardous undertaking. I am certain that their self-reliance, fostered through years of training as a Scout, will help them to meet successfully, the known and unexpected perils of their journey.

For centuries man has dreamed of a trip to the Moon. For the first time, our science, technology, and economy can make this dream come true.

There are still a good many adults who say that it can't be done. And there are others who say that it might be possible, but that it's not worth the effort or the tremendous cost. There are, admittedly, many tough obstacles to the exploration of space. Overcoming them will demand months and years of hard work by thousands of dedicated scientists, engineers, technicians, and other trained workers.

For Project Apollo, America's Manned Lunar Landing Program, we will need a huge, powerful, and reliable rocket. Your father may be working on this project at Marshall Space Flight Center, where the Saturn vehicles are under development. Other NASA Centers are hard at work on the Apollo Spacecraft. Launch facilities are under construction at Cape Canaveral.

The Space Age is truly here, and everyone must accept it, for the results of space exploration will ultimately affect the lives of everyone.

Because man *can* explore space, he *must*. And he must do it as rapidly as his expanding technology and resources permit.

While some reluctant, doubtful adults may have to be dragged screaming across the threshold of space, I have never found any reservations about its exploration among our youth.

That is why I always enjoy talking with youngsters whenever possible. Your confidence and enthusiasm about the future are contagious. I share your optimistic outlook that the Age of Space can be the most fascinating, the most challenging, and the most abundant that the world has ever known.

Thank you for inviting me to be with you today. I hope you will study and work hard so that you can share in the exciting promises of the future.

*** END ***

– Chapter 26 –

Vice President Lyndon B. Johnson wrote von Braun encouraging him to accept a speaking request in New Orleans. The audience was the Judicial Conference of the U.S. Court of Appeals for the Fifth Circuit. "A large group of federal judges and outstanding lawyers"[77] would attend. Johnson was the most consistent and intrepid fighter for a manned Moon landing project, next to von Braun and a few others! His campaign for a strong space program began within hours after the launch of Sputnik on October 4, 1957.[78] Von Braun presented the following speech, May 30, 1963.

THE UNITED STATES EFFORT TO EXPLORE SPACE IS A TRULY NATIONAL UNDERTAKING that will demand the best of all of us. Our objective, as stated by President Kennedy, is to achieve a preeminence in space second to none, and to use the skills and knowledge thus gained for the benefit of all our people and those of other countries.

The weighty responsibility for attaining this goal has been given the National Aeronautics and Space Administration, among the youngest, and certainly the fastest growing, federal agency in the land. Since NASA was created in 1958 its budget has increased 17 times over—from $339 million dollars to a proposed budget of $5,712 million for fiscal year 1964. Never before in our peacetime history has a government agency grown so rapidly.

The NASA program for space exploration has been outlined to cover a broad spectrum, and has been organized in depth in areas of advanced research and technology, space sciences, applications, and manned space flight. About two-thirds of NASA's budget will go into the manned space flight program. As you know, one of the major goals in this program is a manned lunar landing within this decade.

For centuries imaginative men have dreamed of a lunar voyage. But the Moon was considered by all practical people the exclusive property of poets, young lovers, dreamers and mad scientists.

Now, for the first time, man's science, technology and economy are able to support his imagination of the past. Man's dreams of a trip to the Moon will soon become a reality.

In spite of the advances made during the first five years of the Space Age, folks are asking me, "Why do you want to go to the Moon? Why don't you just stay at home and watch television, as God intended for you to do?"

No explanation should be needed except the satisfaction of man's innate curiosity. Man was born to look under the big rock, to climb over the next hilltop, to peer into the anatomy of the atom, and to fasten his gaze on the stars. History has proven that it pays—often in the most unexpected manner—to satisfy this inquisitiveness. But this answer is not usually sufficient for the sponsor of research or exploration. For instance, Columbus secured backing for his famous voyage of discovery by pleading better trade relations with China. While we must admit that his declared objective has not yet been reached, just look at the fallout from his trip.

Our primary reason for going to the Moon, then, is to obtain valuable scientific information relating to a clearer understanding of the universe. Because the Moon has no atmosphere—no wind or water to cause erosion—its surface has changed little in billions of years. The history of creation has been preserved on its surface for much longer period than we can find in the changing face of the Earth. The Moon has also been impacted by particles from outer space which would have burned up in the Earth's atmosphere. What we find on the Moon may help us to unravel many mysteries of the universe.

A trip to the Moon brings into sharp focus all the reasons for accelerating our space program in the first place. It has become a national objective, firmly announced by the President, and clearly understood by the public. Success is essential to our prestige in the role of world leadership that has been thrust upon us by events of recent years. The manned lunar landing will require essentially the same progress in science and technology that will be needed to reach our broader objective—preeminence in space.

NASA has accepted the challenge for accomplishing this remarkable feat, but we are not underestimating the difficulty of the task. We are faced with numerous—but not insurmountable—problems of design, engineering, reliability, checkout, and launching of launch vehicle and spacecraft. In addition to developing the needed hardware, we have the vital problem of keeping three men in healthy and active condition while they are traveling through an environment for which they have simply not been created. Radiation, meteoroids, and solar flares constitute hazards to their well-being.

And then there is the Moon itself, about which we know very little. Scientists have advanced many novel and conflicting theories recently about the lunar surface. Different theories, resulting from ground observations made at a distance of some 240,000 miles, are that the Moon's surface has a texture as solid as granite, is like rock foam or dusty cobwebs, or has fuzz an inch thick.

Some speculate that our spacecraft will bog down or sink completely out of sight. The truth of the matter is that we just don't really know. But we intend to find out before our first astronaut climbs out of his spacecraft. Gathering information and paving the way for a lunar landing program and other manned missions into space have been the Ranger Moon probes, the Mariner Venus probes, and numbers of scientific satellites that have sent back to Earth invaluable information about the space around us.

The problems seem so numerous and overriding to some folks that they throw up their hands and say, "It can't be done." Well, they said Robert Fulton would never propel a ship by hitching a paddle wheel to a steam boiler; and that Henry Ford was foolish for trying to put a gasoline engine on wheels; and that the Wright Brothers were downright crazy for putting wings on one.

Folks in our own time have said that planes would never break the sound barrier; and that an object placed in space would never survive the "heat barrier" encountered on reentry into the atmosphere. Today they are saying that an astronaut can't pass through the radiation belts surrounding the Earth and live; or that his spacecraft will be punctured by tiny meteoroids speeding through space at fantastic speeds. And even if he does adjust to a weightless state for prolonged periods, a solar flare will get him.

Well, you know the answer. Man will surmount these obstacles to space travel just as he has overcome countless other problems during his slow and difficult progress upward from caveman to spaceman.

During the remainder of my talk, I would like to outline NASA's manned space flight program, and discuss in some detail the role of the Marshall Space Flight Center in providing the Saturn launch vehicles for Project Apollo, our Manned Lunar Landing Program.

The Manned Space Flight Program, as you know, includes three major projects: Mercury, Gemini, and Apollo.

In Project Mercury we have taken the first step in the manned exploration of space, have determined man's capability to live and work in a space environment, and have developed the foundation for the technology on manned space flight.

In Project Gemini, with its two-man spacecraft, we will gain operational proficiency in manned flight, and will develop new techniques, including rendezvous in Earth orbit.

In Project Apollo, with its three-man spacecraft, our objective is to establish United States preeminence in space and to develop the ability and equipment to explore the Moon.

The first flights in Project Mercury were suborbital. Alan Shepard, the first American in space, rode the Mercury spacecraft after being launched by a modified Redstone missile, provided by the Marshall Space Flight Center. So did Gus Grissom. Glenn, Carpenter, Schirra and Cooper were launched atop the larger Atlas vehicle. In the Mercury flights the time under conditions of zero gravity was extended from a five-minute weightlessness period in the first 15-minute flight, to the 34-hour, 22 orbit flight of astronaut L. Gordon Cooper.

The Mercury flights demonstrated the soundness of our manned space flight concepts. We have learned many things from project Mercury, but most of all, we have learned that man can contribute materially to the exploration of space. Gordon Cooper brilliantly demonstrated that man can enhance the system's reliability through his capabilities as a test pilot and engineer. And he can act as a scientific observer in space. These demonstrations of man's ability to judge, to reason, and to cope with the unexpected, encourage us to go ahead with the belief that man can be an explorer in space just as he has been an explorer on the surface of the Earth.

Project Gemini has been under way for more than a year and a half. The first unmanned capsule may possibly be launched by the end of this year, with the first manned flight before the end of the next calendar year. The basic objectives here are to increase our operational proficiency and our knowledge of the technology required for manned space flight capabilities. Time in orbit will be extended up to two weeks.

The Titan II missile will be used as the Gemini launch vehicle. The Atlas/Agena will be used as a Gemini target vehicle in the rendezvous and docking experiments. The Gemini program will contribute importantly to Project Apollo in the major phases of a rendezvous maneuver: launch-on-time, maneuvers in space, acquisition in space, and docking.

The Moon will be explored in Project Apollo. This program, directed at the Washington level by D. Brainerd Holmes, is principally supported by three NASA field centers. The Marshall Center provides the Saturn launch vehicles; the Manned Spacecraft Center at Houston trains the astronauts and provides the spacecraft; and the Launch Operations Center at Cape Canaveral provides the launch facilities and operates NASA's "Moonport."

The Marshall Center was formed in 1960 by the transfer from the U.S. Army to NASA of about 4,400 Civil Service employees and an integrated complex of engineering, laboratory, fabrication, and test facilities then valued at 100 million dollars. Our employees now number about 7,500, and our laboratories have been enlarged through additions and new construction. Our Marshall people have the know-how for conducting a rocket program all the way from origin of the project through design, development, fabrication, and testing—with the indispensable aid, of course, of major contributions from contractors.

Our primary task of the Marshall Center for the next few years is to provide Saturn launch vehicles to support Project Apollo. There are three members of the Saturn family—Saturn I, Saturn IB, and Saturn V.

The Saturn I will test the Command and Service modules of the Apollo spacecraft in Earth orbit. The first four Saturn vehicles, with the first stage only live, have been launched successfully from Cape Canaveral. Each success further demonstrated the soundness of engineering design, and paid tribute to the painstaking efforts of the Saturn team to obtain the maximum in quality assurance and reliability in manufacture, testing, and launching. The Marshall Center is building eight of these boosters for research and development flights.

We are looking forward eagerly to the fifth launch, scheduled for this fall, for it will be a significant milestone in the Saturn I program. Several firsts will be achieved.

The second stage will be live for the first time. This stage is under development by the Douglas Aircraft Corporation at Sacramento. Its engines burn the exotic liquid hydrogen/liquid oxygen fuel combination, which gives us 40 percent more thrust per pound of propellant than the conventional oxygen/kerosene combination used in the first stage.

The fifth launch will be the first one from our new facilities that make up Launch Complex 37, our second Saturn launch area at the Cape. All previous Saturn I launches have been accomplished at Launch Complex 34.

The eight engines in the first stage have generated 165,000 pounds each during the first four flights. This fall they will be flight tested for the first time at their full-rated capacity of 188,000 pounds each. The booster will carry a full propellant load of 850,000 pounds, and will be programmed to burn for about 2½ minutes.

The second stage will also carry a full propellant load of about 100,000 pounds, and will burn for more than seven minutes. The second stage will inject itself, a guidance and instrumentation module, and a dummy payload into Earth orbit. The entire weight in orbit will be about 35,000 pounds. No object this heavy has yet been injected into orbit from the Earth in a single launch.

After the Saturn I will come the Saturn IB. This will be a composite vehicle, formed by placing a third stage from Saturn V atop a Saturn I first stage. This will give us the ability to place about 16 tons into Earth orbit, as compared with 11 tons for the Saturn I. With the Saturn IB we will be able to test the entire Apollo spacecraft, partially fueled, in Earth orbit. These manned flights will concentrate on crew training and module maneuvering and docking.

The largest launch vehicle now under active development is the Saturn V, our Moon rocket. It will be able to place 120 tons into Earth orbit, or boost 45 tons to escape velocity.

Let me illustrate the weightlifting capabilities of these three launch vehicles by a comparison. The Mercury spacecraft in which our astronauts have ridden weighs about 3,000 pounds. The Saturn I will be able to place the equivalent weight of seven Gordon Cooper capsules into Earth orbit; the Saturn IB will lift the equivalent of eleven Cooper capsules; while the Saturn V will be able to lift a payload equal in weight to eighty Mercury spacecraft.

In addition to the development of these three launch vehicles, the Marshall Center is responsible for related research, and for the conduct of studies which might lead to space transportation systems of the future. It is not surprising, then, that our workload at the Marshall Center has increased terrifically since the President said we should go to the Moon within this decade. We have been adding about 1,000 new employees a year since that announcement, in a planned, orderly growth which should level off in a couple of years.

While we have added more people to our payroll, enlarged our laboratories, and have a booming construction program underway at the Marshall center, we make no claim to carrying the ball alone. Our projects are far too numerous and too tremendous for that. We have always relied heavily upon contractors for major contributions in all phases of our programs, and shall continue to do so in the future. At the same time, we feel that we must do a certain amount of research and development work ourselves, to keep our knowledge up to date and our judgment sharp in the management of our programs. By doing so we feel that we can merit rather than command the respect of our contractors. While our in-house effort is extremely important, like the yeast that leavens the whole loaf, it represents only a small portion of our total budget. In the proposed budget for the next fiscal year, only 8 percent of Marshall's dollars will be spent at home, while 92 percent will be placed through contracts.

When our workload expanded, we selected the Michoud plant in September of 1961 for the fabrication of boosters that will bear the stamp "Made in New Orleans." This is a government-owned, contractor-operated facility. Our general manager is Dr. George N. Constan, who now has about 190 Marshall employees on his staff. There are now a total of about 6,350 employees at Michoud. The Chrysler Corporation has about 2,350 people working on the fabrication and assembly of first stage boosters for the Saturn I. Chrysler's first Saturn I booster will be completed at Michoud before year's end, and will be shipped by barge to the Marshall Center for static firing early next year.

The Boeing Company, responsible for the production of first stage boosters of the giant Saturn V, now has about 3,090 employees at Michoud. Mason-Rust, the supporting services contractor, has about 650 employees on its payroll.

Boeing-built boosters will be static-fired at the first test site now under construction in southern Mississippi. We are acquiring fee simple title to 13,550 acres of land—an area roughly five miles square—for our Mississippi Test Operations [MTO]. The test site is surrounded by a sound buffer zone for about six miles in all directions. Easement rights are necessary for most of the 128,000 acres in the buffer zone. This will allow owners to continue farming, grazing, lumbering and mining operations, although habitable structures cannot be permitted in this area. The Army Corps of Engineers is acting as NASA's agent for land acquisition, in cooperation with the Lands Division of the Department of Justice. Land acquisition and easement costs will total about 16 million dollars.

At present we are using existing dwellings at the Mississippi site for office space. The pink stucco building that formerly housed Shorty's 43 Club is now the reception area for all MTO visitors.

Construction is just getting under way on the first portion of facilities, which will cost about $200 million. Two other increments will be started as soon as the first portion is satisfactorily under way, bringing the total cost of the facilities here to about $500 million.

The water transportation system to be used at the test site calls for improvement of about 15 miles of river channel and the construction of about 15 miles of canals and docks to allow barges to bring rocket stages and engines directly to test stands. Saturn V second stages, built by North American Aviation at Downey, California, will also be brought to Mississippi for acceptance testing.

We static fire each Saturn I booster made at the Marshall Center two or three times before sending it down the Tennessee River by barge for flight tests at the Cape. The people in Huntsville proudly claim that this booster is the world's largest manmade steady-static noise generator. In addition to that, it can fly. But it won't lift the payload we must have to get men to the Moon. So we are working on a still larger noise generator, the S-IC booster for the Saturn V/Apollo.

The five F-1 engines in the Saturn V booster will develop 7.5 million pounds of thrust. This is five times the thrust of the Saturn I booster we are now testing. This does not mean that it will be five times as noisy, however. When we stepped up from testing a single H-1 engine to a cluster of eight, the noise level did not rise proportionately.

The F-1 engine burns liquid oxygen and kerosene. While all five engines are going full blast, this stage consumes more oxygen than the combined populations of North and South America. It gulps 15 tons of propellant per second.

As you know, most sounds are created by vibrating bodies that cause disturbances in the air. A booster generates sound by exhausting rapidly moving hot gases into the atmosphere. For some time out Test Division has conducted an experimental program in suppressing the sound generated during a firing. The power source that generates the high-intensity sound is the velocity of the rocket jet itself. To suppress the sound, you must reduce the velocity of the jet as soon as it leaves the nozzle. The problem is aggravated because the jet is not only traveling at a high velocity of several thousand feet per second, it is also composed of extremely hot gases on the order of 3,000° Fahrenheit.

A possible solution that has emerged from our experiments is the addition of water to the jet. This approach has two advantages: while the water is reducing the velocity of the jet, it also cools the hot gases. If a sound suppresser of this type can be perfected and used, it would by no means muffle the sound altogether—but it might reduce the sound level by as much as 15 decibels. A woman's voice produces about 65 decibels, and a power lawn mower about 115. When a Saturn V booster is static tested—without sound suppression of any kind—we expect it to generate about 100 decibels as far as eight miles away. That is the reason for having a buffer zone around the test site. Since the tests will last for only a few minutes at the most, and since the sound will decreases rapidly as distance from the stand increases, we do not expect tests to be too annoying outside the buffer zone.

We have come a long way in a short time in America's Manned Space Flight Program. We are spending a respectable share of the national resources on a lunar landing effort—not only in dollars, but in skilled personnel, which is perhaps a scarcer commodity. An effort of this proportion would be futile if it were not to be followed by deeper probes and greater proficiency in space.

Exploration beyond the Moon will continue on the evolutionary approach used in Projects Mercury, Gemini and Apollo. The tempo of future programs will be governed by the progress we make on present assignments. Manned space flight programs of the future will be determined by the national need and the availability of resources.

NASA has the responsibility for studying the future and recommending possible courses of action. Obvious candidates for follow-on projects are the establishment of a Moon base for prolonged occupancy and extended exploration of the lunar surface; a manned space station, or Earth satellite, which could serve as a laboratory for space observations; and manned reconnaissance of the planets.

Much longer Earth orbital flights are needed for testing men and equipment in the space environment, to determine whether artificial gravity is necessary for interplanetary travel, and to demonstrate the reliability necessary for distant journeys.

Future projects will require launch vehicles and spacecraft designed specifically for the intended mission. We are already far along in the study of nuclear upper stages for present launch vehicles, and an ionic, or electric, propulsion system for extended voyages deep into space.

Our Manned Space Flight efforts depend on the success of NASA's space science program, which is using unmanned satellites to obtain a wealth of scientific information on the space environment, the Earth, Moon, Sun, planets, and the galaxy. This data is of great value for our understanding of the physical universe and its affect on the design of launch vehicles and spacecraft.

We at NASA are always conscious that our national prestige and security are heavily involved in our peaceful and scientific pursuit of space exploration. Our government's capacity for constructive international leadership and our economic growth are extremely sensitive to our scientific and technological progress. Other nations are continuously passing judgment on the ability of the world's greatest democracy to make decisions, to concentrate effort, and to manage vast and complex technological programs. In a sense, the ability of representative government and the free enterprise system to meet the challenges of our scientific age are being tested in our space age.

Success will demand the best from each one of us.

*** END ***

– CHAPTER 27 –

Acceptance of American Citizen Award at the 11th Annual German-American Day Festival in Bergen, New Jersey.[79]

IN MY ENTIRE CAREER NO HONOR HAS TOUCHED ME MORE DEEPLY THAN YOUR presentation of this American Citizen Award. I accept it with a deep sense of humility, and shall cherish it always. It will be a symbol and constant reminder to me of man's noblest civic ideals.

This Festival has combined for me nostalgic memories of our ancestral homeland with a quickening patriotic fervor for our adopted country. I am proud of America and its heartfelt desire to see all men free. This is our country's true national purpose.

I know that those of you who left Germany during the thirties and forties came here with a passionate desire to escape a system of men above the law. You came, as I did, to a land known to millions as the country of hope, freedom, and opportunity. I have never regretted the decision.

After my arrival here, there followed years of study of the language and the customs of the American way of life—and patient waiting. Then came the day I received my American citizenship. It is like a new birth.

There is always the danger that a newly-naturalized citizen will become so engrossed in his work, and will devote himself so fully to enjoyment of the pleasures of comfort and success, that he seldom pauses to appreciate his citizenship, and does little to preserve the basic freedoms that it represents.

We must never forget that we will enjoy freedom of religion, speech, the press, and peaceful assembly only as long as we are willing to defend them. The greatest service that this 13th Annual German-American Day Festival can render is to arouse us to a more conscious appreciation for America and all that it means to us, and a firmer resolve to speak out about it. No individual, no people, wanting to enjoy freedom can afford complacency in today's troubled world.

Each of us has a responsibility to preserve and extend equality of opportunity, so that the individual may advance according to his ability and his effort. It is a personal responsibility, for this principle has meant much to each of us. We know it works.

I am proud of our country because it accepts the challenge of a new frontier, and is ready to explore, to change, to improve itself. We are now hard at work on a huge national program to explore man's newest frontier—space. Our success in space exploration is inseparably entwined with the future of America. These are twin loves for me. I would like very much to see the America that I love *excel* in the exploration of space, which has always fascinated me. My life is dedicated to that end.

I thank you.

*** END ***

— CHAPTER 28 —

SHAPE (SUPREME HEADQUARTERS ALLIED POWERS EUROPE/NATO)
SHAPE PROGRESS REPORT SHAPEX 63 [80]

GENERAL LEMNITZER, GENERAL STOCKWELL, GENTLEMEN:

During the last few days we have heard a lot about the dreadful problems besetting this world. Thus I hope many of you will feel a sense of relief by listening to a speaker who will talk about the subject of reaching *other* worlds.

For me, the opportunity to talk to you is not only a great honor, and a rare distinction, but it is a scientific challenge as well. For, in the past, in our space launchings we have been aiming at only one star at a time. Today, however, it seems that I am aiming at no less than 465 stars assembled in this room.

GENERAL

My purpose today is to place the United States space program in perspective. My preliminary and concluding remarks will briefly—and I hope clearly—recapitulate what our space program is, discuss some of its important features, and disclose why I believe it is a sound, worthwhile undertaking.

The United States space program has two distinct goals and is organized into two distinct divisions of authority and responsibility. One purpose is to establish and maintain a strong position of national security within the sphere of space activities. The Department of Defense is designated to accomplish that goal. We also want to advance and utilize scientific knowledge and technologies, related to space and space flight, for the benefit of all nations and peoples. The responsibility for accomplishing this belongs to the National Aeronautics and Space Administration.

NASA objectives may be summarized as:

1. The unmanned exploration of interplanetary space with probes and satellites to accumulate all possible information.
2. The manned exploration of the Moon and eventually the planets of the solar system.
3. The assimilation and dissemination of data gathered by manned and unmanned space exploration.

With these things in mind, the United States has committed billions of dollars, a vast amount of natural resources, and a large share of its scientific brain power to its space program.

Since man is still a self-sufficient species on Earth, and since there are portions of his own planet not yet completely explored, some say that it is not absolutely necessary that he explore space at this time; but because man is inherently inquisitive and intrinsically restless and ambitious, he has set out to explore and exploit space as the next logical step in his continuous search for new knowledge and opportunities.

President Kennedy very eloquently expressed our need for a space program when he said, "We set sail on this new sea because there is new knowledge to be gained, and new rights to be won, and they must be won and used for the progress of all people. For space science, like nuclear science and all technology, has no conscience of its own. Whether it will become a force for good or ill depends on men, and only if the United States occupies a position of preeminence can we help decide whether this new ocean will be a sea of peace, or a new, terrifying theater of war."

The United States is concentrating its effort on the scientific exploration of space for peaceful purposes; however, space undeniably affects our national security.

The exploration of space is contributing directly to the technical knowledge needed for the continuous updating and improvement of all branches of the Armed Services. Construction of Earth-based, worldwide communications networks, development of giant space carrier vehicles, perfection of space rendezvous techniques and space navigation, improvement in instrumentation, design and fabrication of space vehicles, and accumulation of more knowledge of space hazards and bioastronautics are all useful for military purposes, if the need should arise.

Unlike the founding fathers of America, who set out with very limited information concerning the new lands, scientists and engineers today are carefully collecting data and preparing solutions to the problems of space journeys. By means of sounding rockets, orbiting satellites, space probes, and manned Earth-orbital programs, many of the risks have been evaluated and diminished, or altogether eliminated.

Concurrently with these lunar, planetary and interplanetary probe programs, we at NASA are actively engaged in a Manned Space Flight Program, whose major objective is to carry man to the Moon. But before we get into this, let's look at our carrier vehicles that are used in this fascinating program of space exploration.

LAUNCH VEHICLES

Since I am in the Saturn space carrier business, I hope you will indulge me if I broadly classify space carrier vehicles, for our purposes here, as pre-Saturn, Saturn, and post-Saturn types. This is simply a convenient way to handle a large and diverse assembly of rockets used for an equally large and diverse number of space missions. In the pre-Saturn category I will place those vehicles based on large, military IRBMs and ICBMs, and a few specially developed carriers such as the Vanguard and Scout. The Saturn series comprises a group of three in varying stages of design or development. The post-Saturn series is largely of a theoretical nature, although we have performed a great deal of study in this area on specific types of vehicles.

Looking at the pre-Saturn class first, we can say historically that it began with the Juno I, which orbited our first Explorer satellite. This early carrier vehicle was developed from military hardware available at the time. We used an elongated propellant tank and engine from the Redstone ballistic missile, and a cluster of 11 scaled-down Sergeant solid propellant motors for the upper staging. With this vehicle the U.S. Army successfully orbited an 11-kg satellite that is still in orbit.

One other early space carrier that deserves mention is the Juno II, which was developed from a military vehicle. Here we had a carrier vehicle that consisted of a first stage that was an elongated Jupiter IRBM. The upper staging consisted of the same solid propellant cluster used with the Juno I. Ten of these vehicles were launched, with five of them performing their assigned missions.

More sophisticated was the Vanguard, a space carrier vehicle designed as such. While the results of the 11 vehicles launched may seem small—only three satellites were orbited—we learned a great deal about space carrier vehicles technology from that project.

Today we still rely on the modified military vehicle for our smaller space carrier vehicles. Typical of these are the Thor-based vehicles and the Atlas-based series. The Thor IRBM formed the basis for three different carriers: the Thor-Agena B, the Thor-Delta, and the Thor-Epsilon. The Thor-Agena B is a two-stage vehicle employing a liquid propellant second stage. The Thor-Delta is a three-stage vehicle using a liquid propellant second stage and a solid propellant third stage (actually the third stage Vanguard motor). The Thor-Epsilon is a two-stage vehicle with a longer-burning liquid propellant second stage than the other two. These Thor carriers have been the backbone of our scientific payload investigations, orbiting such Earth satellites as the Pioneers, Explorers, Tiros, Discoverers, Telstar, Syncom, Transit, Courier, and Anna.

The Atlas-based series is also one of our mainstays in the scientific satellite and interplanetary probe program. The Atlas-Agena B consists of the military Atlas, less its upper staging for warhead, etc., and a liquid propellant second stage (the same Agena stage as used with Thor-Agena B). With this vehicle we have orbited a good many military satellites as well as placed the Ranger scientific spacecraft into Earth orbit, onto the Moon, and into orbits about the Sun. It also sent the Mariner 2 interplanetary probe past Venus and into orbit about the Sun last August.

We also have under development a vehicle sometimes called the Atlas-Centaur and sometimes simply the Centaur. It consists of the Atlas first stage, with slightly uprated engines, and a newly-developed second stage employing a liquid hydrogen/liquid oxygen engine. This vehicle will be available for our second generation scientific probes such as Surveyor and Voyager.

Another Atlas-based carrier has just finished its service career in a brilliant fashion: it placed Gordon Cooper into his 22-orbit trip around the Earth on May 15. The Atlas-Mercury had the basic Atlas first stage with uprated engines and carried the Mercury manned capsule instead of a second stage.

Before leaving the modified military vehicles, I would like to mention briefly two that are coming up: the Titan 2 and Titan 3. The Titan 2 will be available for our Gemini manned spacecraft program and consists of a Titan liquid propellant first stage and a second stage also using hypergolic liquid propellants. The Titan 3 will have a Titan 2 with an additional high-energy, liquid propellant engine and two solid propellant motors attached in parallel. These three-meter-diameter solid propellant motors will give the basic vehicle an additional 900,000 kg of thrust.

Now to the Saturn or heavyweight class.[81]

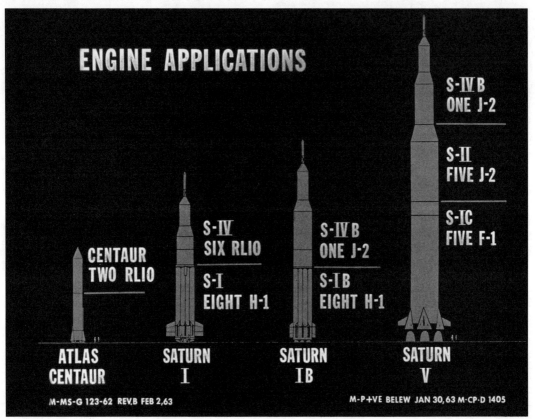

Figure 6 – Atlas, Saturn I, IB, and V

Currently we are working on three versions of the Saturn: the I, IB, and the V. The first of these has already been test flown with a live first stage on four occasions. The second and third are still in the design stage.

Saturn I is a two-stage vehicle. It is composed of known and reliable components and is produced by a technology that has steadily developed through 35 years experience. The first stage of the Saturn I has a cluster of nine propellant tanks, which are manifolded together and which feed a cluster of eight engines. The eight outer tanks are each 1.8 meters in diameter, while the inner tank is 2.5 meters in diameter. The whole stage is 6.5 meters in diameter and 25 meters tall. Each of the eight engines produces over 85,000 kilograms of thrust, for a total of 680,000 kilograms. The propellants used are liquid oxygen and a high grade of kerosene.

The second stage for the Saturn I consists of a single propellant tank with an ellipsoidal dome separating it into two compartments, the upper for the liquid hydrogen fuel and the lower for the liquid oxygen oxidizer. Special insulated pipes from the liquid hydrogen tank run through the

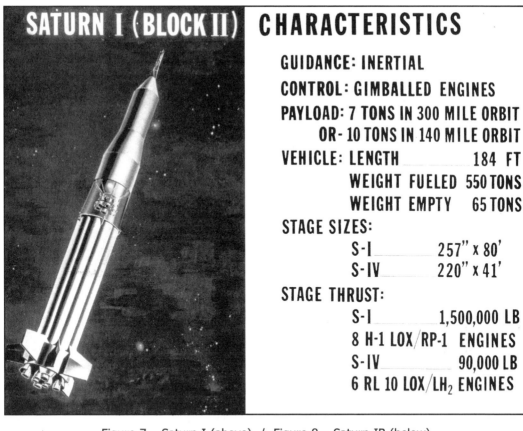

Figure 7 – Saturn I (above) / Figure 8 – Saturn IB (below)

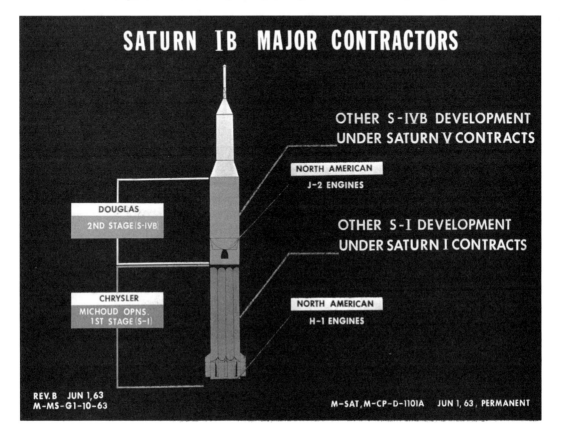

oxidizer tank to the fuel pumps on the engines, which, like the first stage, are also clustered. Six RL-10 engines, each producing 6800 kg of thrust, give a total of 4,800 kg. The first live flight test of this stage is scheduled for later this year when, incidentally, it will place itself into Earth orbit. Weighing nearly 16,000 kilograms, it will be the largest manmade object placed into orbit to date.

Our space mission for Saturn I is to orbit two of the modules of the Apollo spacecraft about the Earth to prove them out, as well as to train the astronauts in some of the intricate maneuvers they will be called upon to perform.

A variation of the Saturn I is the Saturn IB. Here we take the third stage of the Saturn V vehicle and place it atop a slightly modified Saturn I first stage. With this combination we can place into Earth orbit 14,5000 kg, or the entire Apollo spacecraft. It gives us a means of testing out the Apollo craft in an Earth orbit before committing it to a cislunar space. The overall height of the Saturn IB is 50 meters.

Figure 9 – Saturn V

Still in the development stage is Saturn V—the greatest of the Saturn family. With Saturn I and IB we have been talking of space carrier vehicles with thrusts great enough for Earth orbital missions; but for escape from Earth we need a vehicle that can accelerate a payload to some 40,000 km per hour. To do this we have designed Saturn V. Its first stage has a cluster of engines like the other Saturn first stages—but the propellant tankage is different. Here we have two separate tanks, one above the other. The propellants are the same: liquid oxygen and high-grade kerosene, but the engines are entirely new. There are five F-1 engines, each of which produces over 680,000 kg of thrust, as much as the entire first stage of Saturn I. Together they total 3,400,000 kg. This stage is 10 meters in diameter and 42 meters tall. It will have four fins around the base to aid in aerodynamic stability.

The second stage of the Saturn V also uses the clustered engine concept, but it has a single tank with an integral dome separating it into two compartments, one for the liquid oxygen oxidizer and one for the fuel, which in this stage is liquid hydrogen. A single J-2 engine produces 91,000 kg of thrust, and the cluster of five that propel this stage give it a total thrust of 455,000 kg. The stage is 10 meters in diameter and 25 meters long.

For a third stage the Saturn V has a unit with only one J-2 engine and a thrust of 91,000 kg. This stage is 6.5 meters in diameter and is 18 meters long. With the Apollo spacecraft in place atop the third stage the Saturn V will be 106 meters tall, or about the height of a 35 story building.

The mission for the Saturn V is, of course, to place the Apollo spacecraft on its lunar trajectory.

The huge Saturn V measuring, more than 350 feet, will stand taller than the Statue of Liberty. It will be able to lift the same payload as 25 Boeing 707 jet aircraft combined. It can toss 120 tons—more than the weight of one complete 880 aircraft—into orbit 300 miles above the Earth. Or it can hurl 45 tons to the vicinity of the Moon.

I am often asked, "Why do you want to go to the Moon?" One little old lady once asked me, "Why don't you just stay at home and watch television, as the good Lord intended?"

I must admit that I, like numerous others in the space program, am compelled by an irresistible curiosity to explore the vastness of the universe. Outer space has three dimensions—each one of which stretches to infinity. During a slow and sometimes painful struggle upward, man has progressed from cave man to space man, and is able for the first time to break the fetters that have bound him to Earth, and to venture into space.

Our first extraterrestrial journey will be to the Moon. The President of the United States has openly announced our national goal of accomplishing a manned lunar landing—hopefully within this decade. It is an objective that is universally understood, against which our progress can easily be evaluated. The organization for accomplishing this goal has been formed, the method adopted, and preliminary steps firmly scheduled. The cost of the program has been estimated at $20 billion.

However, before I get deeper into the subject of manned lunar flights, let me just spend a few minutes on our almost equally intriguing unmanned orbital activities.

Unmanned satellites of the Earth are giving us a tremendous knowledge of our planet. Unmanned satellite studies have fallen generally into four categories: meteorological, communications, geophysical, and astronomical.

Weather has been one of man's worst enemies since his first day on Earth. The ability to forecast weather accurately is of tremendous importance. Each year billions of dollars and thousands of lives are lost because of man's [in]ability to forecast the weather.

Weather forecasting has made great advances in the last hundred years, starting with the invention of rapid communications in the 1840s and the international meetings of meteorologists that began in 1853. Even though weather forecasting is now a science, much is still to be done.

The United States has already made a major breakthrough in this area with the use of meteorological satellites. The objectives of the NASA meteorological satellites are:

1. Development of satellite system equipment and techniques for an improved understanding of the atmosphere and the development of an operational meteorological satellite system.
2. Cooperation with the U.S. Weather Bureau in the establishment and support of a national operational meteorological satellite system.

The Tiros project is our first attempt to use satellites in weather forecasting. With six successful launches out of six attempts to date, the project has been very successful. For the first time, on November 28, 1960, the U.S. Weather Bureau distributed data from cloud pictures taken by the Tiros II.

Each of our manned orbital flights has relied on Tiros for weather forecasts. For this reason, *Tiros VI* is giving accurate weather information for astronaut Schirra's flight [Mercury-Atlas 8].

The Nimbus project is the follow-up to Tiros and its (Nimbus') objectives are similar. Nimbus was initiated as a NASA research and development project, early in 1960, and basic preliminary design was evolved by NASA's Goddard Space Flight Center. Work on the actual hardware began late in 1960 and early 1961. The first prototype is scheduled for launch in the fall of this year.

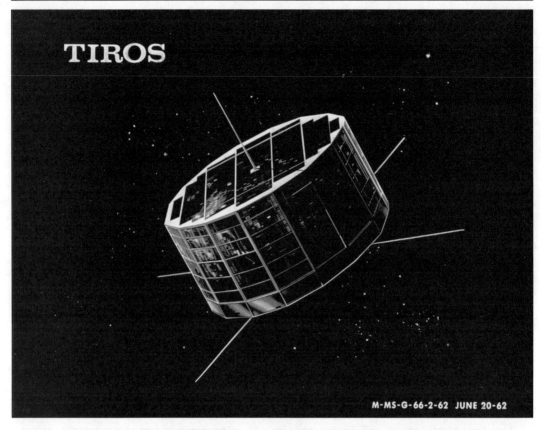

Figure 10 – Tiros satellite (above) / Figure 11 – Nimbus satellite (below)

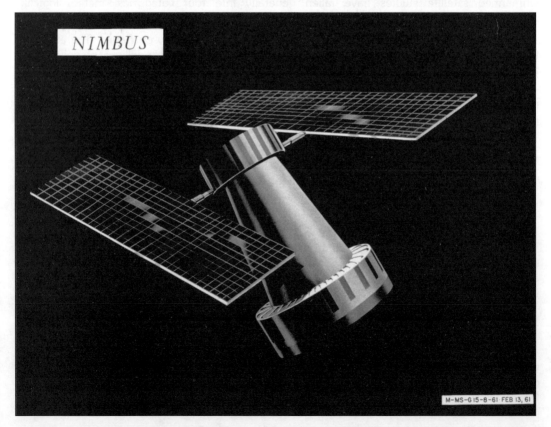

The major advantage Nimbus has over Tiros is its orbit. Nimbus is to have a near-Polar orbit, giving it complete coverage of the world every 24 hours. Tiros sees only about 20 percent of the Earth daily. Nimbus will take pictures simultaneously over an area 2400 km wide and 644 km along the flight path, which will give a much larger field of view than Tiros. The automatic slow-scan television system will transmit pictures every 208 seconds. It will also be equipped with radiation sensors for night photography.

Nimbus' first concern will be in producing cloud cover maps, which weathermen consider "nature's own weather map." Later Nimbus satellites will carry equipment to sense such weather factors as: (1) temperature at various heights, (2) the differences in thermal radiation and reflected solar radiation, (3) differences in radiation absorbed by the Earth and atmosphere, and (4) the distribution of moisture. Eventually, the Nimbus system will consist of a number of satellites orbiting at once.

Figure 12 – Telstar

In addition to weather satellites, we have also had a great success with communications satellites. The spectacular success of Telstar, launched in July last year, impressed the world with the potential of the communications satellite. Although the Telstar experiments have provided a wealth of knowledge, there are still problems to be solved before a reliable system can become operational.

NASA's communication satellite program is based on (1) insuring full development and realization of communications satellite potentials through continued research, development, and flight testing, and (2) assisting in the early establishment of operational communication satellite systems.

We will use three basic means in accomplishing these goals: active satellites in intermediate orbits; active satellites in 24-hour synchronous orbits; and passive reflector satellites in low orbits.

Figure 13 – Syncom (above) / Figure 14 – OAO: Orbiting Astronomical Satellite (below)

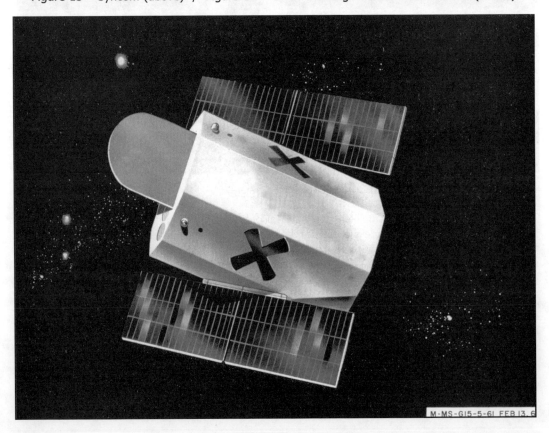

Telstar and Relay are examples of active satellites in low-Earth orbit. Telstar was the first satellite built with private funds and placed into orbit. We continue to learn a great deal about the technology of communications satellites from them.

Syncom is an example of an active satellite in 24-hour synchronous orbit at an altitude of 35,700 km. Although our first attempt failed because of electronic difficulty with the satellite, we did prove that a synchronous orbit could be achieved.

General objectives for the research and development program of new space communications systems are:

1. Intermediate altitude, active repeater satellites with more communication capabilities, multiple access, passive or semi-passive control systems, and powered by solar cell battery combinations, or by nuclear isotope power supplies.
2. Stationary, fully-stabilized, high-gain satellites with long orbital lifetimes, stationkeeping, and attitude control provided by electrical thrusters with solar and nuclear power supplies.

Geophysical and astronomical satellites provide us a means of scientific investigation that can be made with spacecraft in the vicinity of the Earth. They study the light from the stars that cannot penetrate the Earth's atmosphere, radiation from the Sun and its effect on the Earth's radiation belt, the aurora, and the atmosphere. They also measure magnetic fields in space and the flux of energetic particles.

The study of the stars is done with a telescope in a satellite such as the Orbiting Astronomical Observatory (OAO). The purpose of this type of satellite is to collect observations outside the distortion of the Earth's atmosphere. OAO will weigh 1500 kg, of which 450 kg will be experiments. The experiments will be different with each OAO, depending on the questions to be answered. It will be capable of tracking a star with an accuracy of one tenth of a second of arc. This is equivalent to pointing a telescope at a basketball 800 km away. The satellite will orbit at 800 km altitude at a speed of 29,000 km per hour, and will circle the Earth in approximately 100 minutes.

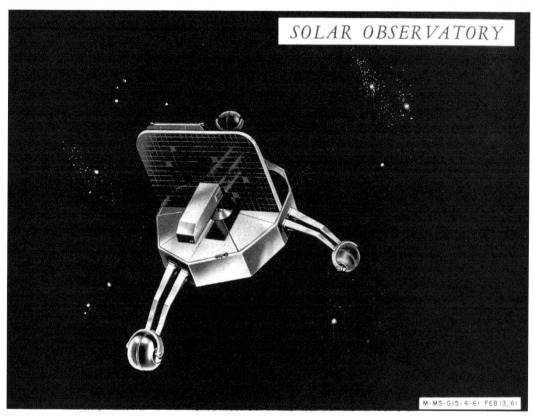

Figure 15 – OSO: Orbiting Solar Observatory

An Orbiting Solar Observatory was launched March 7, 1962, to study the spectrum and the intensity of solar ultraviolet rays and solar gamma rays. Present plans call for two launchings per year for the next five years. The present study of the Sun is especially important because of immediate scientific gains which will support our man-in-space program.

It is impossible to predict the results of such studies. A few years ago some scientists began to wonder, "What keeps the Sun hot?" Their investigations led to the discovery of thermonuclear reaction, and thus to the application of thermonuclear power.

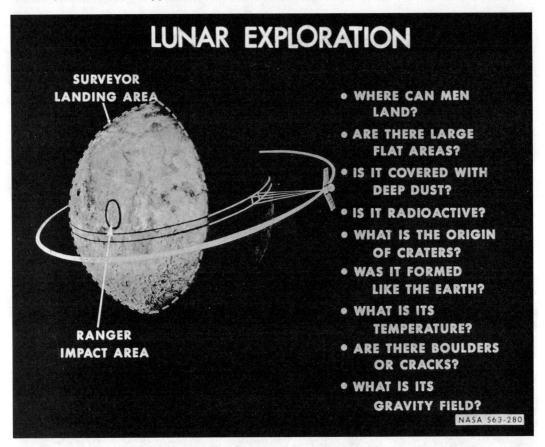

Figure 16 – Lunar Exploration

Unmanned Space Probes

From satellites, let's move briefly to unmanned scientific deep-space probes. We have three types of missions for our space probes: lunar exploration, planetary exploration, and exploration of interplanetary space and the Sun. The first, lunar exploration, is currently being approached through three programs, using the Ranger and Surveyor space probes.

The Ranger program was begun in 1960 and will probably continue through 1965. Objectives of the program include hard-landing an instrument capsule on the Moon, televising pictures back to Earth, and making scientific measurements of lunar environment. There have been five Ranger shots, all of which were launched with an Atlas-Agena B. The first two were Earth orbits, designed to proof-test equipment, and of the next three, two missed the Moon and one impacted on the far side at 9,600 km per hour, destroying itself. Although these shots did not complete their missions, they did transmit valuable scientific data regarding gamma-ray intensity on the Moon and did permit testing of attitude and guidance systems.

Following the Ranger series is the Surveyor lander series, to be launched by an Atlas-Centaur. The Surveyor landers are designed to land on the Moon with the use of retrorockets. This will permit the survival of the spacecraft and its instruments.

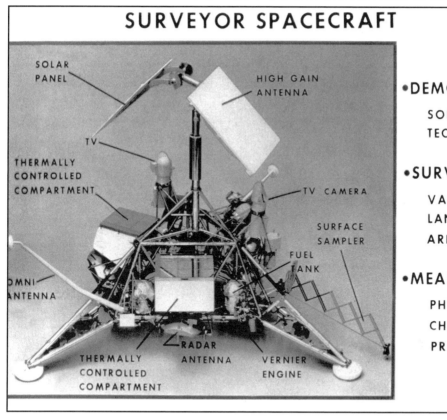

Figure 17 – Surveyor Lander (above) / Figure – 18 Mariner II (below)

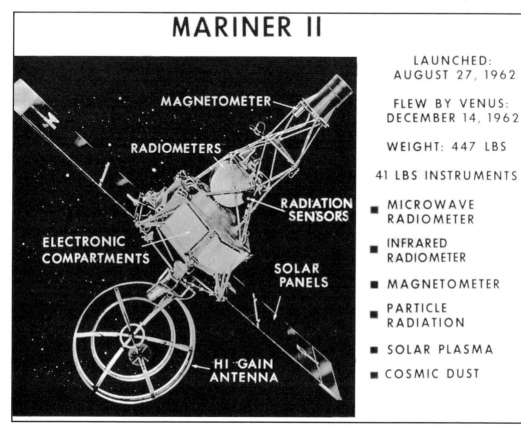

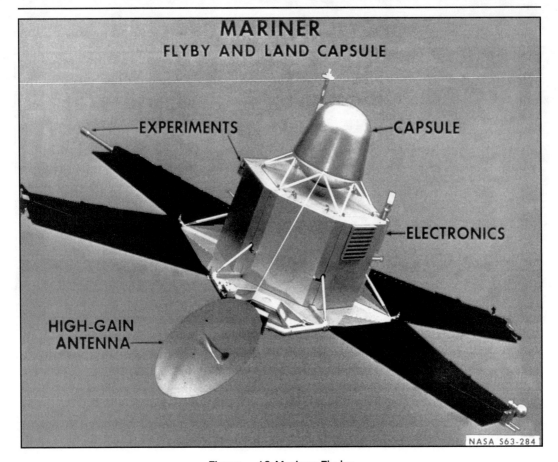

Figure – 19 Mariner Fly-by

Our first planetary probe was *Mariner II* in August 1962. Placed in Venusian trajectory from a "parking" orbit around the Earth, the spacecraft passed within 33,800 km of Venus on December 14, 1962. The probe performed exceedingly well, and a great quantity of useful scientific data was obtained.

Mariner spacecraft are being prepared for fly-by missions to Mars during the 1964 opportunity for flight to that planet. Beginning in 1965, larger Mariners will be launched, using the Atlas-Centaur. Current plans are for an advanced Venus mission in 1965 and an advanced Mars mission in 1966. It is planned that these Centaur-class Mariners will be used on subsequent planetary missions extending at least through 1967, and probably longer.

Manned Orbital Flight

I have touched briefly on the unmanned side of space exploration; now let me move on to man in space.

With the successful recovery of Mercury capsule *Faith 7* and Astronaut Gordon Cooper on May 15, we brought to a close the first phase of our Manned Space Flight Program. Cooper's flight was the sixth perfect launch out of six tries with the Mercury-Atlas vehicle, the last four of which were manned. This perfect score in manned launchings still leaves us a little awed. But it wasn't all luck—Project Mercury was a carefully conceived and developed program. In our philosophy of space flight, man maintains a function of command in the man-machine complex. The Mercury flights demonstrated that man can be depended upon to operate the spacecraft and its systems when it is desired that he do so. Without the presence of a trained pilot in a command function, several of the Mercury missions might have ended in failure.

As John Glenn said, Cooper's flight further demonstrated that "man can be plugged right into the circuit," making unnecessary much of the redundant automatic systems that were designed into the spacecraft for unmanned flight.

The Mercury spacecraft was a bell-shaped capsule, 9½ feet in height, and with a base diameter of 6 feet. This was puny by Russian standards, but it supported Cooper in space for 34 hours, and could have been used for much longer missions.

When Cooper stepped from his capsule, he experienced a slight dizziness when he first stood. His pulse rate was higher, there was a significant drop in blood pressure, and a noticeable reddening of the legs and feet occurred as blood pooled in his lower extremities. This condition continued for several hours, but disappeared to a large degree after he took a nap. Schirra experienced similar symptoms. NASA doctors have concluded that all astronauts may be subject to this orthostatic hypotension—caused by relaxation of the circulatory system during weightlessness—during and after reentry. They are studying ways to alleviate this circulatory disorder.

In summary, it can be stated that all physiological responses during the Mercury orbital flights were in the acceptable ranges, and that the condition of weightlessness and the stresses of space flight were well tolerated. But we still do not know enough about the effects of prolonged weightlessness.

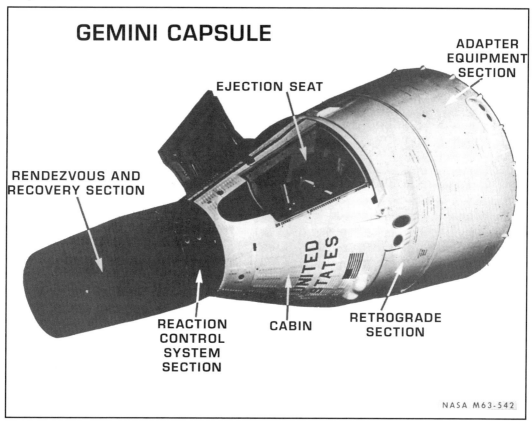

Figure – 20 Gemini

Project Gemini is the next step in our Manned Space Flight Program. Gemini flights will expose our astronauts to progressively longer periods of weightlessness, from about two days to two weeks. Beginning with the first manned launch in this program, scheduled for next year, the two-man crew will face increasing responsibilities as pilots.

The purpose of the Gemini program is to demonstrate orbital maneuvering and rendezvous, extravehicular activity by an astronaut, and long-term flight. The two-man capsule will have the ability of landing under control of its astronauts on a preselected area on land rather than the sea. Briefly the program calls for the 3200 kg Gemini capsule in orbit about the Earth to rendezvous and dock with an Atlas-Agena D stage also in orbit about the Earth. During some flights, the program calls for depressurization of the capsule, while one of the astronauts leaves it, with a 30-

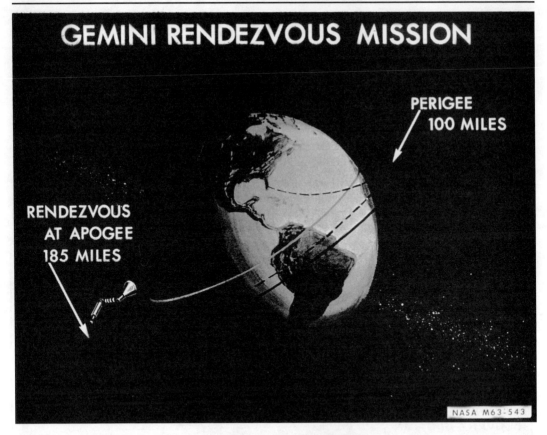

Figure 21 – Gemini Rendezvous

minute supply of air strapped to his space suit. From this experiment we will learn much about the extent to which astronauts can maneuver in space and perform maintenance tasks on their ships.

The main chore for the Gemini astronauts will be to rendezvous with another object and dock with it.

Here's the way they will go about it. An Atlas space carrier vehicle will launch the Gemini capsule into orbit. While both satellites are in orbit, the ground control computers and guidance system will initiate the signals to perform the maneuvers necessary to bring the two capsules into a proximity that will permit the astronauts in the Gemini to perform the final docking maneuvers.

The program will begin with a series of unmanned flights to qualify both the spacecraft and its carrier vehicle. These will be followed by long-term orbital flights with astronauts aboard. During these flights the astronauts will practice terminal maneuvers by means of a small target carried in the adapter section of their Gemini capsule and ejected from it. The final phase of the program will be manned orbital flights and the development of the skill and techniques for rendezvous, using the Agena

Figure 22 – Atlas

D for a target. In later flights of this phase, the astronauts may perform post-rendezvous maneuvers using the reserve power capability of the Agena D, including controlling its attitude and re-igniting its engine.

Hopefully, we will begin our first Gemini flights in the fall of next year.

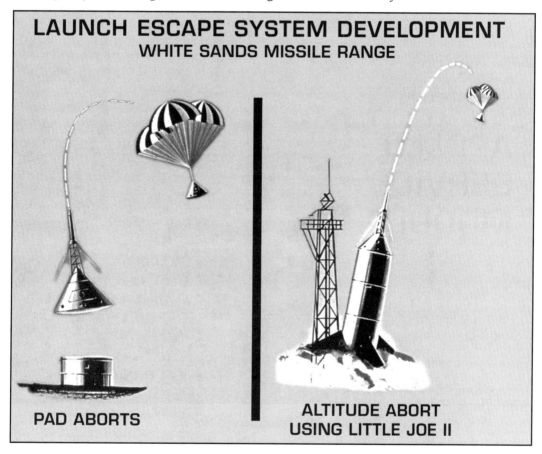

Figure 23 – Apollo Launch Escape System

MANNED SPACESHIP ACTIVITIES

From the Gemini we will proceed to the Apollo three-man space vehicle. Hardware models have been built and tested, and actual components of the capsule are now being checked. With Apollo, we will place man on the Moon, the final phase of our currently authorized Manned Space Flight Program.

The testing of propulsion systems for all of the Apollo modules will be carried out at the White Sands Range in New Mexico, where a special spacecraft propulsion development facility is being constructed. Other tests at White Sands will include a simulated pad-abort escape maneuver and a high-altitude test of the launch escape system.

The Apollo manned space capsule consists of three modules or component parts, which are detachable. These are the service module, the command module, and the lunar excursion module. The service module contains the propulsion system and other equipment not directly needed in the other two modules. The command module is what might be termed the operations room of the Apollo. In this module the three astronauts will stay during their voyage to the vicinity of the Moon. It contains the life support system and the control apparatus. The lunar excursion module, as the name implies, is the small vehicle in which two of the astronauts will actually land on the Moon from the orbiting Apollo command and service module.

The maneuvers by which we get the Apollo to the Moon and back are intricate, but not unsolvable. The Saturn V space carrier vehicle will place the 3-module Apollo spacecraft and its

Figure 24 – Apollo Service Module

third stage into a "parking" orbit about the Earth. After checkout of equipment and crew, the third stage propulsion system will re-ignite, boosting the spacecraft to escape velocity. Once on a lunar trajectory, the command and service modules will detach themselves from the third stage; and the service module will turn about and dock with the service module.

At this point the third stage will be jettisoned, and the Apollo spacecraft will continue to the vicinity of the Moon, a trip of some 72 hours.

Near the Moon, the rocket motor of the service module will fire briefly, allowing the Apollo to swing into orbit about the Moon, at an altitude of about 95 km. Here the spacecraft will be checked out and the decision made as to whether or not to continue the mission. Assuming everything is in order, two of the astronauts will leave the command module and enter the lunar excursion module (or as we call it, the "Bug").

After the Bug has been checked out, its motor will fire for about 30 seconds to place it into an approach orbit that has the same period of revolution as the command module. This orbit will take the Bug down to 16 km above the lunar service.

At this 16 km point, the Bug's engine will fire again to cancel some of the 6400 km per hour speed it will have with respect to the Moon's surface. As it descends to the surface, the capability of hovering at an altitude of 100 meters and maneuvering laterally for as much as 330 meters will allow the astronauts to select the most desirable landing point. Incidentally, the final touchdown velocity of the craft will be approximately 11 km per hour.

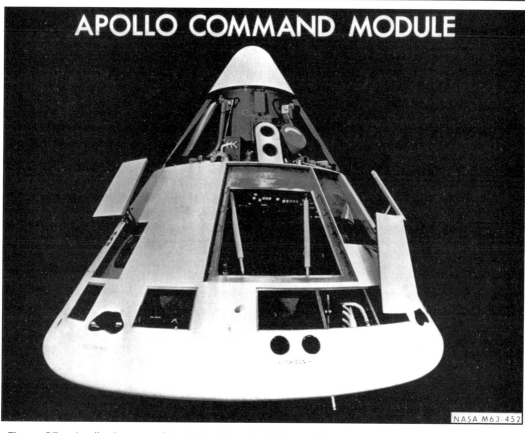

Figure 25 – Apollo Command Module (above) / Figure 26 – Lunar Excursion Module (below)

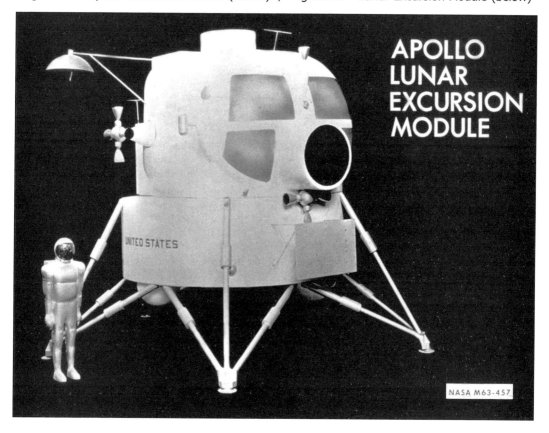

The two astronauts will remain on the lunar surface for 24 hours. One of the astronauts will remain in the Bug at all times. It is planned that they will work in four-hour shifts, exploring the vicinity, collecting specimens, taking notes, and possibly implanting experiments to be left behind.

After a good night's sleep—or perhaps I should more properly say eight hours of sleep—the astronauts will begin the countdown for launch from the Moon. Most of the maneuvering for rendezvous and docking with the orbiting command module will be done by radar and computers. When the two units are a few hundred feet apart, the astronauts in the Bug will take over the manual control and dock with the command module. In an emergency the docking can be done by the astronaut in the command module.

Once docking has been accomplished, the two astronauts in the Bug will transfer to the command module, and the Bug will be jettisoned and left in lunar orbit. The service module engine will then fire its 10,000 kg thrust engine for about 2½ minutes. This will provide the 3200 km per hour velocity needed to escape the lunar gravitational field and to place the service module/command module on the proper Earth reentry corridor. Once this is accomplished, the service module is jettisoned.

During reentry, the offset center of gravity of the command module will permit maneuvering. Once the main aerodynamic deceleration has taken place, a system of parachutes will be deployed to lower the capsule to the Earth's surface, impacting with a velocity of some 7 meters per second.

Looking beyond our Apollo program into the longer-range planning, we are concentrating on three areas, namely, orbiting space stations, lunar-based operations, and interplanetary manned and unmanned missions.

Orbiting space stations are being studied by several of our contractors. One proposed design is a non-rotating station that could be launched by a Saturn IB and used as an interim configuration to provide information on the design of larger space stations of extended lifetime and zero G capability. An extended lifetime space station is also under study. A large multi-radial-module configuration space station could provide facilities for scientific experiments at several levels below 1 G. Resupply of personnel and materials is a vital consideration in any such space station that would remain in orbit and have no reentry and landing capabilities. A logistics spacecraft launched from ground bases will be needed to ferry replacement stores and personnel.

Launch and rendezvous techniques for the support of the space station will be similar to those developed for the Gemini and Apollo programs. Methods of docking, cargo transfer, and stowage of the logistics spacecraft are now under study. Our present blunt-body type spacecraft will afford sufficient functional capabilities and versatility for early phase space station operations, and winged-configuration craft with improved operational flexibility will be available for later, more advance operations.

One problem of a lunar base is the lack of breathable atmosphere. We are now working on an atmosphere renewal system similar to that used in our nuclear submarines. With very limited weight and space, a satisfactory atmosphere has been provided for submerged activity over a period of more than two months.

The three fundamentals—air, water, and food—can be provided for a small lunar colony for a period of one month using present technology and equipment. Housing will probably be provided by flexible domes made of thin tough plastic. Research now offers plastic tougher than the Mylar used for Echo satellites. Confirmation is being sought by unmanned probes as to the nature of the Moon surface. If, as many of our prominent scientists believe, the surface is a labyrinth of caves, then the problems of protecting colonists from the Sun's unfiltered ultraviolet rays and cosmic rays will be greatly diminished.

The major manned objective beyond the Moon is the planet Mars. Perhaps the most formal and detailed approach to manned flight to Mars was made early this year by our Marshall Center and its associated contractors in the contributions to the EMPIRE program. This name was derived from the words "Early Manned Planetary and Interplanetary Round-trip Expeditions," and its purpose is to investigate the problems inherent in undertaking manned flight to Mars and Venus.

Figure 27 – Mars-Bound Spacecraft

The results of one EMPIRE study recommended that nuclear energy be used to propel a Mars-bound spaceship from orbit around the Earth to Mars and return, with chemical propulsion bringing the spaceship components into Earth orbit. Assuming a crew of eight men and initial payload of 45,000 kg, the orbital departure weight for a spaceship would be between 1200 and 1400 metric tons. Total mission times run around 400-450 days.

A typical mission profile for a Mars mission would start with assembly of a spaceship in orbit around the Earth, to be followed by orbital launch and entry onto an Earth-Mars transfer trajectory. Upon nearing Mars, the spaceship would apply retro-thrust to permit entry into orbit around the planet. From the main spaceship a small landing module would be dispatched; it would make an aerodynamic descent and settle on the surface.

After initial exploration, the landing module would be made ready to return to the spaceship waiting in orbit. With apologies to Admiral Anderson, and with a request for General LeMay to note: regarding our landing module aircraft, we have no plans at this time to build a battleship or carriers for Martian canals. Once the crew is aboard, return to Earth can be made. We expect that a convey of two, three, or more spaceships will make the round trip, rather than a single vehicle. Some could be manned, others unmanned, containing cargo, space equipment, and the like.

Beyond Mars and Venus lie many other worlds of the solar system, which we may call astronautical objectives. They include one inner planet, Mercury; some of the asteroids; and the outer planets and their moons.

We cannot yet predict how and when the entire solar system will be explored, but we can say with certainty that man, being what he is, will never be content simply to observe these many worlds through telescopes. Where his eyes can see and his instrumented probes can travel, he will surely follow.

Estimate of the USSR Progress

So far I have been talking about what *we* are doing in space. Now let's look at what *they* are doing.

Contrary to popular belief, the Soviet space program is not a hastily assembled affair consisting of three-fourths propaganda and one-fourth brute-force technology. As early as 1951, the Soviets were launching small animals in sounding rockets and recovering them. The first Soviet Earth satellite was preceded by more than 100 vertical sounding rocket flights. These rockets and their instruments were by and large purely Russian products, although the Russians are shrewd enough to exploit any technology—eastern or western—that will accomplish their own objectives, as they freely admit. The steady progression in capability and skill is evident in the first three Sputniks, with respective weights of 83.6 kilograms, 508.3 kilograms, and 1327 kilograms, and respective perigee altitudes of 228 kilometers, 225 kilometers, and 226 kilometers. These latter figures indicate a highly accurate guidance and control system, considering the time frame.

Once satisfied that they had the basic techniques of space carrier vehicle and satellite instrumentation within their technology, they proceeded rapidly to launch their Moon probes, the Luniks. The first—*Lunik I*—was launched January 2, 1959, while the other two were launched within five weeks of each other in the fall of the same year. Out of these probes came a truly magnificent scientific triumph: the first photographs of the hidden side of the Moon. While they also landed the first Earth object on the Moon, this contributed little to science since it was merely a metal propaganda device inscribed with the hammer and sickle. But the photography of the backside of the Moon was no mere trick; it required the development of a highly precise system of photography and telemetry. For some unknown reason, they waited until April of this year for *Lunik 4*, and we know now that it came a cropper. Specifically, what it was designed to do, we do not know. Possibly it was intended as a soft-lander or was designed to take close-up photographs of the lunar surface.

Turning to Soviet space progress with interplanetary probes, we find that they have launched instrumented probes to both Venus and Mars. Here again they have suffered setbacks. Their failures in this area were revealed by James Webb, the Administrator of NASA, before a Congressional committee. But they have succeeded in getting a probe a long way towards Mars before losing contact with it at one hundred and six million kilometers. Their first Venus probe was launched from a parking orbit in February 1962, marking the first time that this technique had been used in space flight.

Their manned space flight program consisted of two phases, just as ours did. There was the development of the hardware and the training of the cosmonaut—a simultaneous procedure. The Soviet training program was approximately the same as ours, with the exception of the inclusion of parachute jumping. Early clues to the approaching manned orbital flight came early—and the signs were clear. *Sputnik 2*, in 1957, carried a small dog that lived for several days in orbit, although no attempt was made to recover it. In January 1960, they test-fired their large multi-stage carrier vehicle from Tyuratam near the Aralsk Sea into the Pacific. Following this, in May they launched *Sputnik 4*, a pressurized cabin with an anthropomorphic dummy aboard. Here there was a slight setback: the cabin separated from the last stage, but it failed to come out of orbit. In July there were two more carrier vehicle tests into the Pacific from a launching site probably located at Tyuratam, and a month later *Sputnik 5* was orbited carrying two dogs. After the eighteenth orbit, the satellite reentered the Earth's atmosphere on command and began its descent. At 8000 meters, the two dogs in the pressurized cabin were catapulted from the satellite and landed by parachute—a technique used by the second Soviet cosmonaut, Major Gherman Titov.

Then followed the flights of *Sputniks 6*, *9*, and *10*—all testing improved vehicle structures and life support systems, and since weight was no real problem because of the powerful carrier vehicles, the Soviets also included scientific experiments in these satellites as well.

The climax came on April 12, 1961, with the 1-hour and 48 minute orbit of Major Yuri Gagarin, the first man to circle Earth in a spaceship. Again, this was no Soviet propaganda trick. Gagarin wasn't placed into orbit by Pravda and Radio Moscow. He got there the same way that John Glenn and our American astronauts did—he was part of a carefully conceived and developed program of space flight that shows a high order of competence in science and technology.

In March of last year, the Soviets began their Cosmos series of Earth satellites. To date they have launched 18 of them, some from the military missile range at Kapustin Yar near Volgograd and some from the space flight center at Tyuratam. Seven of these are believed to have the weight and configuration of the Vostok manned space capsule. Others are obviously smaller—the ones launched from Kapustin Yar. It is felt that these are primarily military satellites, while the larger ones are forerunners of a new manned space flight attempt. The prediction is not an idle one. On May 15 of this year, the Soviets began another series of space carrier vehicle trial flights, probably from Tyuratam to the Pacific Ocean. The pattern is clear.

I have said before, as others have, that the Russians jumped ahead of us in space—and have stayed ahead in some significant achievements by taking advantage of the greater payload capabilities their high-thrust carrier vehicles provide. I will also second the often-heard motion that they will demonstrate even greater skills, such as the orbital rendezvous of two spacecraft or the soft landing of a robot on the Moon. But even if they do achieve these or other space milestones—some of which they surely will—don't sell our space program short.

For a long time now the USSR has had that same superior payload capability that made it possible for them to attempt some missions on a scale we could not consider. The point is that it was feasible for them to undertake more ambitious programs than we. In fact, they had the basic capability to do more significant things in space than they have done.

Though we acknowledge our tardy start, our program is now sound. It is ambitious—yet realistic—because it is based on sound engineering practices and solid research. With the Saturn payload capabilities, we will have the opportunity to challenge Russia's leadership in the space field. In other words, our biggest disadvantage is about to be overcome, at last making us qualified contenders for all future honors.

As I mentioned before, progressive objectives are built into our long-range goals. These objectives are flexible, as they must be to adjust to the rapid changes occasioned by advances in astronautics. Even though our current schedules are based on extrapolation of our present state-of-the-art and probable growth rates, we consider them sound and firm.

Before closing, I would like to leave with you an aspect of our space program little considered by that segment of our citizenry which conceives of it only in terms of space spectaculars. We have a program based on the free dissemination of knowledge gained in space exploration. Here the Soviets can't begin to compare.

*** END ***

– Chapter 29 –

Fourth National Conference on the Peaceful Uses of Space [82]
Boston, Massachusetts. April 29, 1964

It is my privilege to address the Fourth National Conference on the Peaceful Uses of Space. At the same time, I approach this distinguished audience with great trepidation. And with good reason. My boss, George Mueller, is here. My former boss, Brainerd Holmes, is here, too. If that isn't bad enough, there are many others present who know as much about the space program as I do. But despite these distracting influences, I am going to make this speech anyway, because—and I am sure this is news to no one—the subject of space is very dear to my heart. It is dear to my heart not solely because I am in the launch vehicle business, and it serves my vested interest best to speak out as eloquently as I can in favor of it. The space program is also dear to my heart because I am an American citizen, and I truly believe that this program serves the best interests of the United States, both domestically and internationally.

Space is a popular subject. It is popular not only to those of us directly involved in the various phases of the program, but it is a popular subject with the man in the street. I do not believe there is any national program of any nation which captures the imagination of its people, and the people of the whole world, for that matter, as its space program. Today the names of various spacecraft and astronauts are household words, and are known around the globe. One reason for this unremitting popularity is that, for the first time, man is invading the heavens, a realm which since the beginning of time had generally been left to the vagaries of mystery, superstition and religious speculation. A second reason is the heroism in a spectacular medium on the part of the astronaut. Heroism is a universal virtue, regardless of tongue, regardless of ideology. A third reason for this worldwide popularity is that any space feat, regardless of the nation that sponsors it, is looked upon by people throughout the world as mankind's assault on the unknown. It represents man's effort to conquer his environment and man's effort to understand the basic forces of nature. It is for these reasons that accomplishments in space have had such universal propagandic value. I personally do not feel that the propaganda impact of significant space events on the peoples of the world has been overrated.

Every man harbors deep within him certain visions and dreams that he wishes could come true. I would like to share one of my personal dreams with you tonight. I look forward to the day when mankind will join hands and face the heavens in solid phalanx to apply the combined technological ingenuity of all nations to the exploration and utilization of outer space for peaceful purposes. I applaud the efforts of the President of the United States—Lyndon Johnson today and John F. Kennedy before him—to encourage all nations to work together in the great adventure that is just beginning. Steps taken to date have been comparatively meager, but at least we have made a start. Would it not be ironical—as well as instructive—if nations first learn to transcend their national interests many, many miles away from Mother Earth?

This is but the dream. The realities of today's world sober us to the fact that our technological utopia in outer space has not arrived, and indeed may be a long way off. But I am convinced that the objectives of the National Conference on the Peaceful Uses of Space cannot be achieved until the scientists of all nations can work together in an atmosphere of mutual trust and unfettered cooperation. I believe the nation should continue to work toward this goal without compromising its security, without sacrificing the best interests of its citizens.

I would like to return now to this harsh world of reality and stay there for the balance of the evening. In so doing, my combative instinct immediately becomes aroused because I want to discuss some of the conceptions, or rather misconceptions, about our space program prevailing among certain groups throughout the country. I considered bringing some presentation aids—not slides or films, but loud fire-crackers—to assist me in exploding some of these gross misconceptions.

The first misconception I would like to assault is the idea that the sole mission of the civilian space program is to put a man on the Moon. In the face of the multifarious mission accomplishments of our satellites to date, it is astonishing that this misconception has been able to survive, much less be as prevalent as it is.

The variety and extent of the peaceful space activities of the United States are well-known to all of you and I need not catalog them here. Since 1958 when our first spacecraft, *Explorer 1*, was launched, this country has embarked on a very broad-based space program. We have experienced a variety of spectacular space feats, and I should like to mention a few merely to make the point that the civilian space program is not a one-shot venture. We all followed *Mariner II* making the 36-million-mile trip to Venus, passing within 21,000 miles of the planet and radioing back to Earth important scientific information on its findings. Some of us in this room have enjoyed personally the benefits of other satellites, Telstar and Relay, by seeing clear trans-Atlantic TV broadcasts and hearing telephone conversations. We all know of the Syncom Satellite which travels in a synchronous orbit and introduces the era of the continuous worldwide satellite communication system. We have all benefited from the weather information provided by the Tiros weather satellite. We are all familiar with the orbiting solar observatory, which this very minute continues to provide knowledge about the emission of energy from the Sun.

Then there is the Manned Space Flight Program. The Mercury Program has been successfully completed. Six Mercury capsules, each containing an astronaut, have been successfully launched and returned to the Earth. Names such as Glenn, Grissom, Shepard, Carpenter, Schirra, and Cooper have entered the lexicon of the nation's heroes. Mercury accomplished its primary objective. It has demonstrated the ability of man to survive in space. It has proven that man is not a liability in outer space, but an asset, and that he can perform useful tasks in a space environment.

Coming up next in the manned space flight effort are the Gemini and Apollo Programs. Extensive efforts are under way in both programs. Gemini will demonstrate that man can function in the space environment for prolonged periods of time. He will learn to maneuver his spacecraft, and to meet and physically join with other spacecraft in flight.

The Apollo Program is even more ambitious. The Apollo spacecraft will be able to remain in orbit around the Earth for periods up to two months. It is the Apollo spacecraft which, after its performance is thoroughly proven in Earth orbit, will accomplish man's first landing on the Moon.

The manned programs suggest another question which has often been posed: What are manned spacecraft going to be able to do in the future, both in terms of peaceful and military missions, after the manned lunar landing has been accomplished? This question can be answered today only with another question: Who knows? We simply are in no position to make predictions here because our experience with men in space is so very, very meager. We have logged only fifty-three hours of space travel thus far, hardly enough to base predictions on anything more than pure conjecture. The only way we can answer this is to expose a lot of people to a lot of travel time in outer space, and then apply what these people have learned first hand in their new environment.

This, of course, is not a novel approach. The modern concepts of air power were not developed in "think factories." These concepts evolved from the practical experiences of the brave young members of the Lafayette Escadrille and other flying groups in World War I, who actually took to the air and tried out such things as synchronized propellers, formation flying, instrument flying, and aerial photography.

And so it will be with manned space flight. As our astronauts log additional hours, we shall learn many things from their experiences, still unknown, that will enable men to perform feats in the space environment that as yet have not even occurred to the mission planner back here in our Earthbound think factories.

From the few random examples which I have listed of space achievements to date, it is obvious that our space program is moving forward on a very broad front. My purpose in stressing this fact, which I know is apparent to everyone here this evening, is to meet head on the rather loose language one hears around the country, language which equates the "NASA Program" with the Moon Program, language which constantly refers to the NASA appropriation as the "5-billion-dollar-man-on-the-Moon" budget. The Apollo project is NASA's largest project, but the story does not end there. Far from it. The Program upon which NASA has embarked for the Peaceful Exploration and Uses of Space is the most versatile space program employed by any nation on Earth. I think the nation should be aware of this fact and take pride in its accomplishments.

The next proposition I should like to discuss is the one that says we should abandon the space program entirely because we cannot afford it; or, as some would have it, reduce the level of effort to a level we can afford.

It is apparent that a program encompassing such a large variety of complex space activities requires for its accomplishment a major commitment of the nation's resources. Today about one percent of the total income of the United States is devoted to the civilian space effort. In this decade, the United States will invest about 35 billion dollars in its total civilian space program. About 20 billion dollars of this will be devoted to the manned space flight effort.

In terms of manpower, the costs are equally high. Today, about one quarter of a million people, both in government and out, are working in the civilian space program. The bulk of these, about 200,000, are part of the government-industry team for manned space flight.

In terms of facilities, the investment again is high. The space program involves far more than merely building large boosters and spacecraft. It involves capital investment in large engineering companies throughout the United States for fabricating, assembling and testing the systems that comprise the launch vehicles and spacecraft. It requires investment in large environmental chambers, centrifuges and simulators for preparation and training. It demands a worldwide tracking and data acquisition network feeding into an integrated mission control center. It requires a highly sophisticated launch complex, such as the Moon port being created at Cape Kennedy. When completed, these facilities will include some of the most massive and complex ground and engineering installations ever designed.

The question presents itself: Can the United States afford a program of such magnitude in the face of its continuing commitments to other national programs such as defense, agriculture and welfare?

There are those who say that we should cancel this "Moon madness" and divert these space funds into more Earthly projects, such as cancer research, aid to the needy and urban redevelopment. Others say we should continue the program, but at a reduced annual level of effort in deference to these other programs. This latter theory holds that although it may take longer to get to the Moon and, although the total cost of the program will run higher, at the same time we shall be proceeding at a reduced *annual* rate of effort, a rate of effort the country can better afford.

My personal view is that we can afford to invest one percent of our annual gross national product in space. I believe that we can afford to continue to invest four or five percent of our annual federal budget into the civilian space effort. I do not believe that if this budget were cut that any substantial increase would automatically accrue to these other programs, programs which, incidentally, I consider very worthwhile. Based on my own personal experiences before Congressional committees, I do not believe that these annual appropriations are solely the result of fiscal finagling with figures, with funds being taken from this agency and applied to that agency, like some juggling act carried out under a master plan. I believe the approach taken by our elected officials is one in which each program must stand or fall on its own merits, as viewed by the American voter.

I believe the pace of our space program is entirely reasonable. Although the goals are ambitious and the schedules tight, it is not a crash program. We are moving forward vigorously, now that our immediate space goals have been clearly defined. I believe we are moving at a pace the American people expect, now that they have given the program their stamp of approval. There is no harm in setting one's goals high. This is the rigorous life. This is the American tradition.

I could not possibly take leave of you tonight without briefly discussing the question which is probably put to me personally more often than any other. It runs something like this: We agree that the nation should have a space program. We further agree that it should move forward on a broad mission front. But to do these things, why is it necessary to go to the Moon? Why can we not develop [Earth orbit] space, and forget this business about going to the Moon?

I think I can best make my point here by using an example. When Charles Lindbergh made his famous first flight to Paris, I do not believe anyone believed that his sole purpose in going was simply to get to Paris. If going to Paris had been his sole objective, he could have traveled by boat in much greater security and comfort. His purpose was more than personal transportation. His purpose was to demonstrate the feasibility of transoceanic air travel, not to get to Paris, but to fly across the ocean. Now he could have selected a wheat field in Alsace-Lorraine, or perhaps he could have landed in one of the moors in Scotland. But Colonel Lindbergh had the farsightedness to realize that the best way to demonstrate his point to his world audience was to select a target familiar to everyone. Everybody knew where New York was, and everybody knew where Paris was. The history books have recorded the immediate impact of his voyage.

Lindbergh achieved his objective, and today we are using air transoceanic transportation, not only to go to Paris, but to deliver cargo to Copenhagen, mail to Manila, tourists to Tokyo and, on selected occasions, maintain the Berlin airlift.

In the Apollo program, the Moon is our Paris. We have selected a target familiar to everyone. Rather than asking the man on the street to accept the esoteric language of the trade, such as "rendezvous," "docking," and "orbital transfer," in defining the immediate objectives of man in space, the late President Kennedy selected a goal which is entirely familiar to the man on the street: sending men to the Moon before the end of this decade. The fellow next door knows what a man is, where the Moon is, and he knows when this decade is out.

To prepare for this lunar trip, we shall have developed space vehicles with the versatility to perform all of the orbital operations presently envisioned by this or any other nation. After the Moon is conquered, this versatile capability remains for other manned space flight applications, in both near and outer space.

The purpose of the manned space flight program, then, is to build an important national resource, a broad space capability, that will enable the United States to investigate and utilize the environment of space for a long time to come.

It is providing the muscle which will undergird the nation's posture in this newest dimension of national power—outer space.

And I can illustrate this same point by treating it in terms of dollars. In this decade we expect to spend about twenty billion dollars on the Manned Space Flight Program. We consider that about 92 percent of this money, or well over 18 billion, is being and will be used to create permanent capital for the United States. Some of this permanent capital will be measured in terms of new technology, industrial manufacturing complexes, and governmental test and launch sites. But the vast majority of this newly created capital will be the large numbers of highly-trained technical people who will provide the nucleus of talent for the space missions following the lunar landing.

The other eight percent of this twenty billion dollars may be regarded as the consumables, as that part of the program which is used up in the process of developing this new capability. This includes such things as materials used up in ground tests, and the hardware and fuels that are actually launched into space.

I have saved until last the question which intrigues me most: Why invest in space at all? Money aside, is there really any purpose to be served by the space program?

To me, the question, "Why invest in space?" is the same as asking, "Why have an age of science?" Man has been born an insatiably curious creature concerning his natural environment, and I think if there is any lesson man has really learned during the last two thousand years of his violent history here on Earth, is the fact that it seems to pay off handsomely, but often in the most unexpected way, to keep satisfying his curiosity about the world around him. The only restraints upon his satisfying this innate curiosity, now that he has shed the shackles of superstition and myth, have been the lack of the proper tools, such as the microscope, telescope, bathysphere—or spacecraft—to enable him to carry his investigations further and his probes deeper. In today's explosion of technology, man is rapidly developing these tools. He is rapidly developing the capability both to explore the Earth more thoroughly, and to explore the celestial environment that surrounds him. And because he is developing the means, man will follow his natural nosiness and will capitalize on his opportunities to investigate and uncover new phenomena of nature. He will, and should, apply these tools to first-hand observations of the environment of space.

Indeed, this is what we have already set out to do. For the first time, we are now in a position to examine and measure the Sun. For the first time, we stand on the threshold of determining the origin and nature of the solar system. And we have already demonstrated our ability to use this new space environment for practical purposes, such as communications and weather observation.

These are the questions that I wanted to discuss with you.

Time permitting, there are many other noteworthy aspects of our space program that I should have liked to discuss with you tonight. There is the subject of the very beneficial impact the program has had on the American economy, in all sections of the United States. There is the vast subject of program management, and the managerial revolution that has swept the country to find adequate means to marshal the varied, dispersed talents of our government-industry team in massive array to get mammoth projects, such as the Apollo program, accomplished. Literally thousands of private businesses, both large and small, are participating in the Apollo program alone. But I have decided not to discuss management at all, for fear my boss would think I am trying to tell him something.

I also could have discussed the contribution that the civilian space program is making to higher education, and the stimulus it has provided to the research programs of our universities and the training of our young scientists and engineers. I could have discussed what we are doing in NASA to transfer to the industrial sector of our economy the results of "spin-off" of our space-oriented research that may have application as new tools, devices, materials, processes and techniques of benefit to the American consumer in everyday life.

And finally, I could have discussed some of the tangible steps that this country, represented by some of our own high-ranking officials in NASA, notably Dr. Hugh Dryden, have taken with other nations regarding international programs for the peaceful uses of space.

I have avoided all of these subjects because I feel that the basic questions, which I have discussed, command priority attention. It is these questions that reach to the vitals of our entire space effort, and it is these questions that must be answered, if this country is to have an adequate space program, or indeed, if this country is to have a space program at all.

The United States has made monumental strides forward into the space age. At the same time, we have hardly begun. The present phase of our space program brings to my mind an earlier period in the history of the Western World, the period when man first laid the great foundation, both in thought and achievement, for the better world we enjoy today.

I speak of the Renaissance. I speak of the era of Michelangelo and Da Vinci, the era of Shakespeare, Cervantes, Raphael and Rabelais. It was in this period, beginning in the fifteenth century, that man took his first great strides forward to emancipate himself from his environment, when man first undertook the conquest of the planet Earth as a place of human occupation.

The Renaissance is often called the Age of Discovery, the age when men summoned forth their courage and set out on the high seas to explore the four corners of the globe. It was the age of Sir Walter Raleigh and Sir Francis Drake. It was the age of Columbus, Diaz, Pizarro and de Gama.

We have not yet entered the Second Age of Discovery, the exploration of outer space. We are still in the harbor. We are still building and checking out the seaworthiness of our craft. We are still learning the things we need to know about the new medium through which we shall have to travel. The Mercury astronauts were not the explorers. The Mercury astronauts were the test pilots, but they did not leave the harbor of Mother Earth.

But we do stand on the threshold of the Second Age of Discovery. When the craft is ready and the oceans of space are calm, calm because we have learned the new medium and have prepared to sail on it, the new explorers will venture forth. The space-age Columbus and Magellan are presently unknown, but they are sitting somewhere today in a public schoolhouse preparing for an adventure that exceeds the wildest daydream which today distracts them from their books.

These are the beneficiaries of our crude efforts today. Here are the people to whom we shall pass the baton. But the first lap of the race is ours. And we shall not falter.

*** END ***

– Chapter 30 –

French War College: Welcome to Students at MFSC [83]

GENTLEMEN:

I am very glad to see you visit our Marshall Space Flight Center, and I wish to extend to you a very cordial welcome.

I have to admit, unfortunately, that it is something of an effort for me to talk to you in French. It was quite some time ago that I began learning French in the French Gymnasium at Berlin. And at present I haven't much opportunity and time to practice it. In fact, time has become a convenience that I am lacking most. Therefore, to my regret, I also cannot spend much time with you today. We have on our agenda a number of conferences with the NASA astronauts who arrived here last night from Houston.

But at least I want you to know that it always gives me great pleasure and satisfaction to see the genuine interest in space exploration which exists here and abroad. Our NASA program is directed towards several well-established goals:

Scientific research in that space that surrounds us;
Planetary exploration;
Development of a broad space flight capability;
Utilization of satellites as a new instrument for worldwide communications, weather observation and navigation;
And last, but not least, the general advancement of science and technology to maintain a leading role in world economy and politics.

Here at the Marshall Center, we are developing and testing heavy launch vehicles. You will see later some of our manufacturing and engineering laboratories, our guidance laboratory, our test division—briefly, all phases of development which are carried out at our Center.

Our primary mission is the development of the giant Saturn vehicles for the Apollo program. A Saturn vehicle will be used to send a manned Apollo spacecraft to the Moon with enough propellant for its return trip to Earth. The development of the Apollo spacecraft and the training of the astronauts is under the direction of another NASA Center in Houston. Our Kennedy Space Center in Florida is in charge of the actual launching of our space vehicles with their "payloads."

The three members of the Saturn family are Saturn I, Saturn IB, and Saturn V. The smallest, our Saturn I, will be able to place an Apollo craft into Earth orbit. It will be used for flight testing the craft, its life support, communications, and control equipment. The more powerful Saturn IB will enable three astronauts to practice rendezvous maneuvers in Earth orbit. Flight tests with Saturn IB will also include the lunar excursion module, with which the descent to the lunar surface and return flight to the Apollo command module, which remained in a lunar orbit, will later be achieved. Our giant Saturn V will finally be used to place the entire Apollo craft with the lunar excursion module into a lunar trajectory.

So far, we have launched—without any failure—five Saturn I vehicles. Flight tests of the Saturn V are not scheduled until 1967. But you will already see in our laboratories the giant F1 engines, which will give you an idea of the tremendous overall dimensions of the vehicle to be used for a lunar landing.

It gives me great satisfaction to know that France also has started an active space program. NASA's Office of International Programs in Washington is in charge of establishing and maintaining close cooperation with ELDO and ESRO, in which France participates. I am convinced that French scientists and engineers, known in the entire world for their intellectual brilliance and creative imagination, will make decisive contributions to the successful progress of space technology and research.

A new dimension has opened to man's activities. By directing our common efforts towards the limitless universe and pursuing an active space exploration program, I am sure that we shall also successfully contribute to peace on our home planet and the welfare of all people.

I hope you will find your visit here interesting and instructive, and I am extending to you my best wishes for the remainder of your tour through the U.S. and for your future careers.

*** END ***

– Chapter 31 –

The University of Alabama main campus is in Tuscaloosa. In January 1950 the University of Alabama in Huntsville Extension Center opened its doors for evening classes in a local high school. A few months later the von Braun team was transferred to Huntsville from Fort Bliss, Texas. Von Braun pledged his help and persuaded some of his associates to teach some classes. As a result of growing government contracts leading to a population boom, Huntsville citizens started to lay the foundation for future growth. Eight acres were reserved by the City of Huntsville as a first installment for a future university project. Eventually a world class research institution was built.

University of Alabama Huntsville Campaign $750,000 [84]

THANK YOU, BEIRNE, DR. ROSE, GENERAL PERSONS, AND ALL OF YOU OTHER distinguished—and, I trust, affluent—guests and friends.

I notice that my good friend Beirne Spragins, in his usual forthright and candid way, mentioned why we are here tonight. I should like to add my own personal endorsement to what he said. We are here to initiate a campaign to guarantee that our community takes full and complete advantage of the opportunity before us: namely, the opportunity to develop the full potential of our human resources by helping to create another full four-year college here in North Alabama.

How can we best support the establishment of this college?

With money, hard work, and more money.

As you know, the University of Alabama has already set aside $500,000. The people of this area have been asked—they're asking themselves, really—to raise another $750,000. This money will be used to pay for building construction and for equipment.

But far beyond the mere physical facilities that this money will provide, lie the other dividends.

One of the great challenges facing modern society is to promote the enrichment of the intellectual, cultural and economic life of the people. Today this task has become a vast and complex one, reflecting the great advances in technology, modern communications and the many specialized demands of a rapidly changing world. This is particularly obvious in Huntsville.

I am persuaded that we who make our homes in this community believe that this area is destined to continue to grow and become a great and permanent scientific, educational and industrial center. The success of our community in achieving these ambitions rests in a large degree upon the development and growth of our educational opportunities.

Why is education so important?

Because education is to growth and prosperity what fertilizer is to cotton and corn.

The National Aeronautics and Space Administration has a vital interest in strengthening the schools and universities of our country. Although we have no direct public responsibility for the promotion and advancement of education, as such, NASA and the Marshall Space Flight Center feel a great responsibility to organize and conduct space exploration in such a manner as to use the best resources of the country.

This campaign to raise funds for the establishment of a full four-year college program for the University of Alabama Center in Huntsville ties in directly with our national space program.

The Marshall Space Flight Center is particularly interested in the expansion of the University of Alabama here to assist us with the tremendously complicated job of developing the boosters for space exploration.

Your challenging job of leading the drive to obtain funds for our university will have the wholehearted backing of all of our people at Marshall and the entire community, I am sure. We are firmly behind the expansion of the University of Alabama Huntsville campus. The completion of the project will not only provide new employees for our national effort, but will give our team a chance to advance technically in their various fields as well as provide them a chance to broaden their horizons.

Even though the progress that the university has made here to date has been outstanding, the breakthrough that will result from giving undergraduate degrees in Huntsville will be a major advancement for this section of the state. By building the academic climate in Huntsville, you will provide the opportunity for our major contractors and for us to recruit new people in the future and will enable us to hold our present outstanding employees.

It is for these reasons that I applaud the leadership of this community and congratulate the University of Alabama in joining hands to raise funds from private sources to construct and equip a new building as soon as possible.

When it comes time—and today is as good a time as any—for you to make your own contribution, I should like respectfully to ask that you allow your most generous instincts to prevail. You might even let them overwhelm you.

I have heard it said that the true test of generosity is whether you can tip a hat check girl a quarter without wondering if a dime would have been enough.

In putting over this drive, I'm sure Chairman Spragins is expecting the most from you. Not some lesser amount that makes you wonder if you are doing enough. As Beirne told me just a moment ago, "We're asking the limit. We don't want anything skimpy," he said. "We don't want any topless bathing suits or even bikinis. We want full measure in every direction." As someone else once said, "You don't buy gasoline, you buy transportation. You don't buy theater tickets, you buy entertainment. You don't buy spectacles, you buy better vision." And of course, "you don't buy a school house, you buy education."

Benjamin Franklin put it even better: "If a man empties his purse into his head, no man can take it from him."

So I should like to urge each citizen to consider seriously what a four-year college program will mean to the economic and social welfare of our community. And I should like to urge that all of us respond to this opportunity as full and as best we can.

The University of Alabama and Dr. Rose are already doing their part, and doing it well.

We can do no less.

I am confident that this campaign will be a resounding success. And I know that each of us who play a role in it will be proud and happy that he did so.

Thank you

*** END ***

— Chapter 32 —

Butler University

More than 20 honorary degrees were bestowed on von Braun. Among them was Doctor of Science from Butler University, Indianapolis, Indiana (1965).

Freedom to Explore [85]

It is an honor and a privilege for me to be at Butler University tonight. I want to express my sincere appreciation to President Jones and to the Board of Trustees for the honor you have bestowed upon me. The honorary degree of Doctor of Science from this respected institution, which emphasizes the highest quality of educational program, and which has an enviable record of more than 110 years of service to the community and the nation, is an award which I shall cherish always.

I am deeply impressed by the qualifications of the faculty members at Butler University, and I somewhat envy the opportunities they have to stimulate and cultivate a thirst for learning among you students. I have always had the highest regard for a person who bears the title of "professor." And I like to think of this word by its original meaning, as one who openly professes something, one who is willing to stand for his convictions with an almost religious fervor.

A university is more than classrooms, more than textbooks, and more than professors, however. The essential ingredient of a great university is academic freedom. I am not talking about freedom to wear mop-top haircuts, sleep on a pallet in the administration building, or use four-letter words. Here, I am sure, you students have the freedom to inquire, to explore the unknown, the freedom to make a choice, to decide for yourself what is right and what is not—and the freedom to be wrong. This freedom to be wrong, to make mistakes and to learn from them, is one of the most precious freedoms of all.

Any student—or any society—afraid to make mistakes is afraid to take chances, to risk new experiments, to accept new ideas, or to take bold steps into the unknown. The fearful student never develops into a good scholar. He clings safely to familiar ground. He learns the time-tested rules and familiar formulas.

Such a student is receiving training but is not being educated. We must not confuse skills with knowledge. The great teacher does not so burden his students with systems, with procedures, or with techniques that they lose their curiosity, their ability to inquire for themselves. There are no formulas to success that can be studied and memorized like multiplication tables.

There must be thousands of graduates of Butler University who have left this campus during the past century to assume places of leadership in industry, government, education, the armed services and other fields. I would like to salute briefly just one Butler University alumnus, Class of 1941; Congressman Richard Roudebush, from the sixth Indiana district. He has been in Congress since 1960, and is a member of the House Committee on Science and Astronautics and its Manned Space Flight Subcommittee.

I have become pretty well acquainted with Congressman Roudebush, for I have appeared before this committee regularly, usually with hat in hand, asking for some more funds for space exploration. There are very few members of Congress who can equal his knowledge of the nation's vast and highly technical program for the exploration of space. None surpass his dedication, determination, and ability. Indiana is fortunate to have a man of his caliber in the House of Representatives.

And the far-sighted professor does not encourage his students to become such specialists in one field that they neglect to educate the whole man. Of course, we need highly trained individuals in all the specialized areas of learning. But for a long and fruitful career it is essential to build a broad foundation in the basic subjects.

We have a saying in the National Aeronautics and Space Administration that hardly anyone can utilize what he learned in school, anyway, for many of the things we are doing today weren't even invented or dreamed of when we were at the University.

In this age of exploding technology, it is important for the university to teach the student how to learn. Having built a broad foundation of knowledge, he at least knows where to begin. And when he learns the difficult art of adding to his knowledge, of continually updating it, he is on the road to success.

If a student doesn't learn this, especially in the scientific and technological fields, he will be hopelessly obsolete ten years after he gets out of school, no matter how good his grades were.

The candidate for a degree, then, should go to his exercises with his head crammed full of useful knowledge, and with his imagination and curiosity very much alive, eager to inquire and explore the unknown. I want all of you to use your imaginative powers tonight as I talk to you about America's future in space. Space is not an arena for the scientist and engineer alone. It touches every one of you.

The National Aeronautics and Space Administration was created on October 1, 1958 during the administration of President Eisenhower. And the past year has been the busiest and most fruitful this young agency has experienced since it was created. The only other government agency that moved at a faster pace and made more history was the fabulous 89th Congress. For the first time since the Space Age began, the nation is reaching its true capability in space exploration.

The impact of space exploration has been felt on human affairs throughout the world. It has altered our thinking in science, education, government, industry, and religion. No area of human activity has escaped. Our ideas about the universe, the toys of our children, the conversations of diplomats, the dreams of our young people, the fortunes of businessmen, and the pronouncements of church officials reflect the influence of man's entry into space.

The nation is deeply committed to the development of manned space travel to the Moon and to the planets. There is no doubt that this is the greatest, the most ambitious, and the most challenging project mankind has ever undertaken. The number of problems—technical, scientific, organizational, and budgetary—is almost unlimited. I am glad to be able to report to you that we are finding satisfactory solutions to these many problems because universities are continually sending us educated young men and women with the talent, the motivation, the dedication, and the curiosity which have always been the trademarks of superior accomplishments.

By far the most decisive factor in the exploration of space is the human element. It is true that we have electronic computers that can calculate a million times faster than the human brain. We have sensors that are a million times more sensitive than the human senses. We have instruments which are a million times more accurate than human perception. And we have materials that can withstand the almost unbelievable extremes of the unusual environment of space. However, a gigantic project like our space program frequently requires courage to accept responsibility, willingness, and tenacity to bear frustration and disappointment, wisdom of judgment, and imagination to project into the future. These human traits are simply beyond the abilities of electromechanical, automated systems. It may be reassuring for us to consider that man still has the decisive role in our modern technological society, and the human characteristics which have distinguished man as far back as we know human history.

The aspects of national security and international prestige are imperatives that motivate us to explore and utilize the new environment of space. Basic to the program, however, is the scientific urge to explore the unknown, to gather knowledge from the outermost reaches of the Universe. The nation's scientists, individually and collectively, and our universities have responded to the need to increase our basic knowledge to meet the difficult challenges that lie ahead of us.

While the scientific community is playing a leading role in the space effort, the space program in turn is exerting a significant and beneficial influence of science. New approaches to establish disciplines have appeared, and there has been a tremendous increase in interdisciplinary activity. The breadth of the space program involves almost every known science and technology—physics, chemistry, biology, astronomy, geology, geodesy and cartography, to name but a few.

The history of science and technology indicates that a small spring of basic research is often the source of a mighty river of technology. We must therefore continue to fill our reservoir of knowledge with new concepts that can come only from basic research.

Our manned space flight program is designed to capitalize on the presence of men in space. Scientific and technological groups throughout the United States are developing for us experiments that will take advantage of man's abilities to use his senses, to manipulate instruments and experiments, to evaluate what he observes, and to conduct investigations in Earth orbit, in lunar orbit, and on to the surface of the Moon.

You are familiar, I am sure, with the Earth orbital missions now being flown in Project Gemini. After Gemini will come Project Apollo, which will give our astronauts the ability to remain in Earth orbit for extended missions, and to make a journey of a quarter of a million miles into space, land on another heavenly body, and return safely to Earth. These missions will be accomplished with the Saturn launch vehicles, now under development.

*** END ***

— Chapter 33 —

Central Intelligence Agency [86]
Washington, D.C., December 21, 1965

MY PREPARED REMARKS ARE BRIEF. I KNEW I HAD A SPEAKING ENGAGEMENT FOR tonight on my calendar, but this CIA outfit is so secretive, I couldn't find out where I was to speak until Carl Duckett picked me up at NASA Headquarters a few minutes ago.

I have always envied you people in the Central Intelligence Agency on your ability to camouflage your annual budget from the piercing eyes of the Congressmen who examine appropriation bills. But I remember a time when even Admiral Raborn didn't have it so good. That was when he was running the Fleet Ballistic Missile program which led to our fabulous submarine-based Polaris missiles. We in NASA still don't have it so good. So I hope you will share your secret with me before the Congressional hearings get underway on NASA's FY '67 budget requests. You see, many people are only too ready to point out that in space we are only second. So, clearly, we have to try harder, and that costs money.

I have been asked to talk to you about NASA's space efforts, particularly Project Apollo, the program to achieve a manned lunar landing in this decade. I have an uneasy feeling that you men in the CIA's Directorate of Science and Technology already know more about our program than you have admitted. But perhaps I can confirm some of the information you already have. Or—maybe your crystal ball is even better than mine.

The nation's space effort is essentially a research and development effort, based on the same broad-based science and technology that nurture our industrial and military strength. Our competition with the Soviets in this field has elevated space accomplishments to a global yardstick for scientific and technological stature. Our technological prowess is the deciding factor today in keeping the peace, and demonstrated technological and scientific capabilities have become a basic source of national power. Preeminence in the field of space exploits has become important in international relations, and influences our dealings with other nations involving peace and freedom in the world.

It is, therefore, of the utmost importance, for our work, that we know what our competitors in this field are doing. This is almost as important as a bigger budget. But, we are operating under a severe handicap, to say the least, because of the fundamental differences in attitudes and outlook of the people and the government of the United States and that of Russia. The Russians, as you know only too well, reveal very little information in advance about their plans. For that matter, their after-the-fact releases of information have been limited to statements that produced propaganda mileage, or admissions of failures they knew they were unable to hide.

On the other hand, NASA is operating its space program in a goldfish bowl. The world can read and hear about it, and even see live on television some of the more spectacular events as they happen—such as the reception of photographs of the lunar surface from *Ranger IX*, the launch and recovery of Gemini astronauts Schirra and Stafford last week, and the recovery of the *Gemini 7* astronauts, Borman and Lovell, after their 14-day mission.

Such things are fine when successful, but embarrassing when not. And yet, we couldn't—and wouldn't—do it otherwise. We think that running an open show is within itself an important part of the message we are trying to get across about America and the Free World.

Our space program is supported by the public, and we have an obligation to inform them of its progress. In fact, we have a mandate to do so. The National Aeronautics and Space Act of 1958, which created NASA, spells out in clear language our duty to the public. It says: "The Administration, in order to carry out the purpose of this act, shall ... provide for the widest practicable and appropriate dissemination of information concerning its activities and the results thereof."

So, we announce our plans far in advance. Even if we should try occasionally to keep our cards close to our chest on some detail, the news media would howl, for they seem to think, we should let them know not only what we are about to do, but what we are thinking about doing—even before we have reached a decision on a matter. In fact, they are particularly astute about digging up stories on upcoming developments.

And even more so than the news media, the Russians have an uncanny way of knowing what we are planning. And they take the wind out of our sails, or embarrass us, by taking that step themselves a few weeks earlier. An example of this is Ed White's walk in space outside of his Gemini spacecraft. Or the attempts to photograph and soft-land instruments on the Moon. The Russians have never been hesitant about reaping the greatest possible psychological and propaganda benefits from the timing of their space achievements.

So we would like to know more about what the Soviets are up to in space. This, of course, is where you people come in, and I couldn't even start telling you how to collect such information. But with your indulgence, I would like to make a few remarks on techniques that are proving more and more useful in the business world. Managers of big industrial firms have to solve problems on how to run their companies—and make decisions which may have direct bearing on profits and losses for years to come—with a great number of "unknowns" in the equation. This, it seems to me, is where there's great similarity between their problems and yours. To cope with this situation, industrial leaders have developed a "modeling" technique whereby they can protect themselves against the adverse effects of at least some of these unknowns.

For instance, the manager would have a study made and a paper prepared on upcoming negotiations with the union for contract renewal. It might say: "Now, supposing the discussions proceed in *this* direction, here will be the result. If we are successful on this point of arbitration, here will be the result; if we lose this point in the bargaining, this will be the result."

The modeling technique now presents the manager well in advance with a whole slew of situations which he may find himself up against at a later date. By visualizing the potential crises well in advance, and by modeling tentative courses of action for each potential situation, he can be prepared to cope with them when the time comes, and even make his decisions well in advance.

This technique may not be too new to you, for you are engaged in essentially the same crystal ball gazing endeavor—trying to take a few pieces of information, evaluate them, and put together a picture out of a vast number of unknowns.

The modern industrial manager, like you, is faced with uncertainty. For instance, he doesn't know whether his company will win certain contracts or not. He doesn't know how military developments such as those in Vietnam will affect him next year. He doesn't know how his next round of negotiations with labor will turn out—whether there will be a wage increase or a strike. He doesn't know what his competitor is up to, and in what shape he will be next year. But he is still confronted with the problem of providing the best course of action, regardless of what happens, predictable or unforeseen. And he cannot afford to depend on spur-of-the-moment decisions or on courses charted under emotional stress.

A recent survey showed that in the aerospace industry, corporations with such long-range planning activities have usually weathered the storms better than those without computerizing.

Since you have the task of completing a picture with just a few pieces of the puzzle, it might be interesting to use the modeling technique, if you have not already done so, in a form suitably adapted to your kind of problem. For instance, you may find it interesting to ask yourself, "How would I react if I were a Soviet factory manager, operating under a set of likely circumstances, if this or that particular situation develops? Considering all the ground rules, standards, methods of accomplishment, and pitfalls in the Soviet system, what would I do?"

Since much of the built-in logic of the Soviet system is well known, it seems to me that use of the modeling technique might shed some new light on problems, such as how they might run their space program, and suggest some answers on possible courses of action on our part. But now I'll stop carrying coal to Newcastle by trying to make intelligent suggestions to people who make intelligence their business.

I know you followed with interest the flights of *Gemini 7* and *Gemini 6* which have just been completed. Needless to say, we in NASA were elated over the successful rendezvous achieved in orbit, for it demonstrated a technique which must be used on the trip to the Moon in Project Apollo.

Many advances in manned space flight are being made in the Gemini program. Another example is the extravehicular activity, such as Ed White's famous "walk in space" outside of his spacecraft. This was training for the day when astronauts will need to leave the spacecraft in which they were launched from Earth for many reasons. Some examples are: to recover an experiment from an orbiting unmanned spacecraft that is not designed for reentry through Earth's atmosphere; to make repairs in flight; to inspect unfamiliar satellites and perhaps disarm them; to make adjustments during rendezvous and docking maneuvers; and eventually to assist in the assembly or orbiting platforms in space; or to transfer to a completed orbiting laboratory for a six-month tour of duty.

In our post-Apollo space program in NASA, we are studying and defining missions in three basic categories; Earth-orbital, lunar-orbital, and lunar surface. In each of the three groupings, present configurations of Saturn V launch vehicle and Apollo spacecraft hardware will be used, with modifications as necessary for the particular mission.

I expect that the majority of our space missions for some time will be planned for Earth orbit. We shall, of course, continue lunar and planetary exploration, but the space near Earth will receive the most attention. Astronauts and their instruments will be looking down upon the Earth during long-duration orbits. Among the areas which give promise of direct benefits to the average person are: meteorology, oceanography, geology, geography, hydrology, agriculture, forestry, and upper atmospheric physics.

The emphasis during Earth-orbital operations is placed on practical benefits. And the potential, believe me, is unlimited.

For instance, the question of population explosion and food production will get increasing attention in the years to come. At the present rate of population growth, it is estimated that the people of the world will number between six and seven billion in the year 2000 A.D. and 35 years later, in the year 2035, world population will be from 12 to 14 billion. This population growth will occur in a period when our children and our grandchildren will be directly involved. The task of systematically developing the resources of planet Earth is therefore of the utmost importance and urgency.

In order to develop these resources, a much better, up-to-date knowledge of the status quo and of trends is necessary. Satellites could help some organization, such as the UN, keep an eye on the situation. One of the basic rules of management is "First isolate and understand the problem, and then apply action where needed."

*** END ***

– Chapter 34 –

The Industry-Education Council, La Canada, California, was organized to develop channels of communication between education and industry in order to maximize the involvement and support of local businesses. This would lead to the fostering of improved student achievement.

The Challenge of the Century [87]
May 12, 1965

It is always a pleasure to address a dedicated group such as we have assembled tonight, and I am delighted to be here.

When a group voluntarily undertakes such a cause as expressed by this council, my admiration is multiplied many times. Reduction of the time lag between the development of new knowledge and its availability in the classroom, and the introduction of new technology to industry are second in importance only to the space mission itself.

We are learning many new things as we build our hardware for manned space flight. We attempt to catalog each advance in technology, and, if it has an application in industry or in the classroom, we freely publish the information for all interested persons.

I want to congratulate all members of the council and urge that you intensify your efforts and make the results of your studies available throughout educational circles.

I was invited tonight by an old, old friend, Dr. William H. Pickering. When I see Bill these days he almost knocks me down with his swelled out chest over the Ranger Moon pictures. And, I can't really say that I blame him for this. I believe the Ranger mission stands at the top in real space achievements. Hard-landing Russian soft landers notwithstanding.

I hope we can borrow some of Bill's space navigators for our Apollo manned lunar expedition. The JPL people have proven more accurate at a quarter of a million miles than some of our flights here on Earth.

When I first started thinking about my remarks today, I thought I would give you a brief report on the progress being made in the development of the Apollo Saturn launch vehicle family.

But, on second thought, I decided perhaps I better take a seat and get a progress report from California. As you know, the Marshall Space Flight Center is responsible for the three versions of the Saturn—the Saturn I, Saturn IB, and Saturn V.

All upper stages for the three Saturns are being assembled by aerospace firms here in California under Marshall Center contracts. Three of the four different types of Saturn engines are under development by the Rocketdyne Division of North American Aviation, Inc., at Canoga Park, California.

In fact, out of the top ten single contracts in dollar value administered by the Marshall Center, eight are with California companies. These top ten contracts total about three billion dollars, with two billion of the total going to the eight California firms.

That is why I thought I might do better, tonight, to sit and listen and perhaps I might learn how well our Moon schedule is really doing. Seriously, these California firms are to be commended for their fine efforts in Saturn/Apollo hardware production. The S-IV second stage of Saturn I, developed by the Douglas Aircraft Company in Santa Monica, has flown four times from Cape Kennedy with four straight successes.

After the Saturn I, the Saturn IB will be next up at the launch pad, and will be the first Saturn to carry men into space.

Douglas at Santa Monica is also building the S-IVB second stage of the Saturn IB, and we are looking for another great performer here, just as we have in the Saturn I.

The Saturn IB first stage has been static tested at the Marshall Center in Huntsville with very satisfactory performance. We are hoping to see the first Saturn IB vehicle assembled at Cape Kennedy late this year, with the first developmental flight early next year. The Saturn IB can place 18 tons in low Earth orbit. It has a built-in capacity for payload flexibility and is adaptable to a third stage. It can be topped off with payloads 22 feet in diameter without using the hammerhead approach. With the addition of a Centaur as third stage, it can send scientific payloads weighing 9,500 pounds to Mars.

The Saturn IB cannot lift the Apollo spacecraft fully-fueled and instrumented for the Moon expedition, but it can orbit it with sufficient fuel for the Earth-orbital training sessions. And it does so at a vast savings over the cost of the same operations using the Saturn V.

And now, we come to the Saturn V—the Moon rocket, sometimes referred to as the "Big Boy." All three Saturn V stages are in various phases of ground testing.

At Huntsville last month we static fired, for the first time, all five engines of the huge Saturn V booster test stage. Excellent results were obtained from the test, with more than 500 measurements of the booster's performance recorded in our new blockhouse. Although the first test was a short one, lasting six and one-half seconds, we are now conducting longer duration firings. If our hopes for continued success are realized, the Saturn V first stage will be ready and waiting for the lunar mission within this decade.

The Boeing Company of Seattle—another non-California firm—is producing Saturn V booster stages for the Marshall Center.

The S-II second stage of the Saturn V has also begun its static firing phase at the North American test site at Santa Susanna. North American Aviation's Space and Information Division at Downey has the S-II contract. Fabrication of the stage is at Seal Beach.

Now, the Saturn V third stage is a slightly refined version of the Saturn IB second stage, the S-IVB, built by Douglas. The major difference in this stage, as it moves up to the more powerful space rocket, is the necessity for restart capability.

Most of the ground testing conducted for the Saturn IB version will be considered adequate for the Saturn V version, also. Douglas has already shipped the first S-IVB flight stage to Sacramento for testing, which should begin in the near future.

With the Saturn stages moving into ground static test programs, it follows that engine development and engine delivery is also coming along in fine shape. North American's Rocketdyne Division is developing H-1 engines for Saturn I and Saturn IB boosters, the F-1 engine for the Saturn V booster, and the J-2 engine for the S-IVB and S-II stages. The Rocketdyne people are supplying the horsepower that will take our astronauts to the Moon.

The Saturn V will become the "heavy duty truck" in NASA's launch vehicle stable. It can place 140 tons into low orbit, send 45 tons to the Moon, 35 tons to Mars, or hurl two tons clear out of our solar system.

Like the smaller Saturns, the Saturn V was planned with flexibility in mind. After the lunar expedition of Project Apollo, the Saturn V can be outfitted with an unmanned Mars or Venus scientific spacecraft. Or a series of Saturn V launches could be used to assemble in orbit a manned expedition to one of our sister planets.

The Apollo spacecraft, with little modification from its present configuration, could be applied to some extremely interesting orbital missions. After orbit by the Saturn V, the Apollo spacecraft, fully fueled and with long-duration life support capability, could remain in orbit for extended periods and could maneuver itself to almost any desired orbit. The Lunar Excursion Module—designed for the actual Moon landing—could be modified and used in orbit as a utility vehicle to enable the astronauts to make "side trips" and inspect other orbiting objects friendly or unfriendly. The Lunar Excursion Module could leave and return to the mother spacecraft as simply as train cars are coupled and uncoupled.

This free maneuvering in space is an area in which we urgently need experience. From these type flights, new and exciting ideas will emerge. There is nothing that beats practical experience. By the time Project Apollo culminates in a manned landing on the Moon, this nation's astronauts—who so far have logged a total of 63 hours in outer space, including the ten hours flown by Gus Grissom and John Young recently—will have added measurably to their practical space flying experience.

The concept of modern air power was a product of thousands of second lieutenants who learned to fly and who familiarized themselves with the new ocean of air that was the challenge of their day. It was these second lieutenants and these ensigns who dreamed up schemes like synchronizing a machine gun with a propeller and using the airplane as a whole to aim their gun. It was these enthusiastic young men who first tried their hand at formation flight and taking photographs from airplanes to see whether the plane had any use as a reconnaissance tool, and dropping wooden bombs into staked-out targets for practice.

I think we have much the same situation today.

We must get men out there in freely maneuverable craft and we must let them log hour after hour in space. Soon they will become acquainted with their ships and their environment and they will begin to look about, seeking new maneuvers, new ideas, and new concepts.

These flight experiences will help to take the "hex" out of space. Right now the environment of space has a certain amount of mystery, even an air of magic, connected with it. This is caused largely by its unknown elements. We have always been afraid of the unknown. We need to dispel that mystery and the only way to do this is to get better acquainted with it, live in it, and work in it, for extended periods. This is one of the broad objectives of NASA's manned space flight program.

Often, whenever space travel is mentioned, the term "hostile environment of space" crops up in the conversation.

Space is not hostile toward man. Hostility is willful, directed deliberately by someone with the intellect and ability to act. So space is not hostile. It is simply there, following the immutable laws of nature as it has done since creation. Neither is it hospitable. There is danger in space for man— if he ignores its laws. There is danger for man on Earth if he walks off a cliff, stays out in the tropic Sun to long, or attempts to live in the Arctic cold without adequate protection.

But, man has learned to cope with his environment on Earth.

And he will live in the environment of space, as he comes to know its characteristics better, and learns how to use them to his advantage. We cannot reap the full benefits of space flight until we are thoroughly familiar with space, with spacecraft, and with the spacecraft's capabilities in its environment.

We are not pushing forward in the Project Apollo Moon expedition just to be able to say we are the proud owners of a handful of lunar sand. In reaching out to the Moon, we are developing our space flying ability to a high degree. Landing on the Moon is not the primary goal of our manned space flight program. The underlying goals are the really significant ones.

In reaching Paris, Lindbergh proved the feasibility of trans-Atlantic flight and opened a new era in commercial aviation transportation.

In reaching the Moon, we will prove the feasibility of space flight. All the hardware; the technical and scientific manpower; the launching, tracking and communications nets; the test, management, and logistics support will remain with us as much more of a permanent asset than a handful of lunar dust.

These are inescapable benefits to all free peoples and to the prestige of the United States. NASA was chartered for the peaceful exploration of space and directed to stress global cooperation with other interested nations. This was to emphasize our historic role as a peaceful and non-aggressive nation.

When Project Apollo is accomplished, only the name will recede into the history books. The knowledge, ability, and other emerging capabilities will form a solid base on which to proceed toward operational flights to the Moon and back and toward the exploration of the solar system.

This is why the late President Kennedy said that we are "not one man going to the Moon, but an entire nation."

Project Apollo is undoubtedly the greatest scientific and engineering project ever undertaken by man. Apollo far outstrips the construction of the Egyptian pyramids, man's first monumental engineering task. Apollo surpasses the Manhattan Project, the modern-day program that developed the atomic bomb. But unlike the Manhattan Project, Apollo is not scheduled on an around-the-clock, crash basis, with unlimited spending. It is following a soundly conceived, carefully planned, step-by-step approach.

Benefits from the exploration of space are already numerous. One of the outstanding benefits is its stimulus to the American people. America is a nation on the move. Its growth has been paced by the urge to explore and conquer the unknown, and the desire for better living.

The exploration of space is fundamentally a research program, based on man's natural curiosity about himself and the universe about him. He has always wanted to know what is over the next hill, under that big rock, inside the atom, and out in space.

We are in the process if winding the mainspring of Project Apollo. With the continued support of the Administration, Congress, and the American people, the steps to the Moon will be ticked off like clockwork, and we shall keep our date in space.

*** END ***

— Chapter 35 —

Dr. von Braun spoke to the State of Alabama League of Municipalities.[88] Represented were the mayors from throughout the state. A resolution was drafted to thank him for his participation: BE IT RESOLVED that the Alabama League of Municipalities expresses its sincere appreciation to Dr. Wernher von Braun for his splendid and informative remarks to the convention and for his arranging for the edification and enjoyment of all of the delegates of the static firing of the Saturn booster.

The League recognizes the invaluable achievements of Dr. von Braun to both the state of Alabama and to the nation and extends its deepest appreciation for his service and devotion to duty in his chosen field of endeavor.

Adopted in convention assembled at Huntsville, Alabama on this 19th day of April 1966.

WHEN MY GOOD FRIEND [MAYOR] GLENN HEARN ASKED ME TO SPEAK TO YOU tonight, I warned him that many of you may have heard me already, for I spoke to the Alabama League of Municipalities in 1959, I believe.

When he assured me that you had probably forgotten what I had to say at that time, I accepted. But I still feel somewhat like the U.S. Senator who was campaigning for reelection after a six-year term in Washington. Getting out among his constituents again, he came to a little Alabama town, and saw a crowd gathered around the courthouse.

On a platform above the crowd stood a man with a noose about his neck, a preacher, and the sheriff.

"Pete," the preacher asked, "do you have any last words or requests to make?"

"No, I don't think so," Pete said.

The Senator elbowed his way through the crowd and said, "Say, if Pete doesn't have anything to say, would there be any objection if he let me use his time to say a few words to this fine group of people gathered here?"

"Well, I guess that would be all right," the sheriff said.

Whereupon Pete spoke up. "Sheriff, I reckon I do have a last request after all. Would you go ahead and hang me first? I've already heard the Senator speak."

This is an impressive gathering of city officials. But tell me—who is minding the store while you are out of town? Who is listening to irate complaints about potholes in the streets? Who is getting the blame for those traffic tickets? And who is trying to pacify homeowners who object to rezoning their residential area for expansion of the business district?

I am glad that you could escape from your daily harassments to visit with us in Huntsville during this annual convention, and to talk shop with other city officials from all over the state. Your organization is to be commended for its concern, not only with the everyday efficient operation of city government, and town planning, but for its interest in the broad creation of a favorable environment for every phase of human activity and association. This is a noble objective for government at any level. It is also an objective that cannot be achieved solely by one level of government alone. Local, state, and national governments must work closely together to achieve the greatness that can be ours. The opportunities and innovations brought by the rapid growth of knowledge and technology today cut across or sectional boundaries.

City limits, county lines, and state borders serve a useful purpose for separating certain responsibilities and authorities of government, such as road building or property assessments and taxation. But—ignorance, sickness, poverty, and injustice, some of the more intangible and humanitarian areas in which we have come to demand more help from the government, also know no boundaries. Solutions to these problems areas in our society demand the fullest cooperation and understanding between government officials at different levels, and between the government and the governed.

A great deal of the credit for the progress and prosperity which this nation now enjoys should go to numerous hard-working, conscientious, and dedicated public servants. The significance of the part which government plays in our daily lives is reflected in the fact that one out of every five Alabamians working off the farm, in our towns and cities, is working for the government—either federal, state, or local. The government is today the second largest non-agricultural employer in Alabama, second only to manufacturing.

If I had not been out of town for several days already this month, I would like very much to attend your session tomorrow afternoon when the state's gubernatorial candidates speak to you. This is the fifth campaign for the office of governor which I have followed since I came to Huntsville sixteen years ago, and I am keenly interested in it. I plan to make my home in Huntsville and Alabama for many years to come.

As a city official, I know you have a personal interest in the viewpoints of potential governors on local issues. Well, as an employee of the federal government, and as a resident of Huntsville and Alabama, I have a natural interest in their views on cooperation between governments at all levels.

I firmly believe that Alabama has the capacity for greatness in the political, economic, and social life of the nation. If we can be No. 1 in football for two, and maybe three, years in succession, we can be first in education, industry, agriculture, research, space, and a dozen other fields, if we make these our goals.

Municipal officials have long been accustomed to working for bigger and better communities. In the future, it might be advisable to concentrate more on community improvement than on growth.

At the present rate of population growth, it is estimated that the people of the world will number between six and seven billion in the year 2000 A.D. And 35 years later, in the year 2035, world population will be from 12 to 14 billion. This population growth will occur in a period when our children and our grandchildren will be directly involved.

Immediate steps must be taken to drastically increase food production, or large parts of the Earth are headed into a series of disastrous famines. The results undoubtedly would not be limited to misery for the people directly involved, but in all probability would cause additional strife and war in a mass struggle for sheer survival. Compared with its misery, the worst disasters in human history would pale in significance.

The task of systematically developing the resources of planet Earth is therefore of the utmost importance and urgency. In order to develop these resources, a much better, up-to-date knowledge of the status quo and of trends is necessary. The space program could help some organization, such as the UN, to monitor the wise management of Earth's resources. One of the basic rules of management is "first isolate and understand the problem, and then apply action where needed." Manned satellite stations with suitable sensor systems will undoubtedly permit continuous surveillance of Earth in such areas as crop planting schedules, harvest results, floods and draught, hydrology and soil erosion, snowfall and subsequent water management of storage lakes, meteorology, ore an oil deposits, oceanography and life patterns in the ocean, population census taking, development of roads and railroads as well as sea and airlanes and a host of others.

Many of the characteristics of the Earth's surface and atmosphere, such as cloud patterns, are best seen or measured from a distance.

Population taking should be relatively easy, for man changes the face of the Earth wherever he is. He cuts trees, tills the soil, and builds houses, factories and roads. These signs of man's activities could be observed and correlated with Earth-based sampling techniques to give a highly accurate quick look count of the world's population.

The causes of crop diseases, water or mineral imbalance in the soil, frostbite, or sunscald may be detected from space. Black stain rust, one of the most damaging of crop diseases, is difficult to detect in its early stages. Remote sensors can spot the rust several days earlier than a man walking through the fields. And in some instances techniques may be devised to detect diseases in crops as much as a year earlier than present ground-based methods, and poor soil conditions that will not yield healthy crops may be detected before the crops are planted. Excess salinity of the soil in cotton fields of Texas as a result of irrigation showed clearly from photographs made in the Gemini program.

To offset the growing consumption of surface water, underground rivers can be detected by measuring the tiny differences in soil temperatures above them. Such streams hold thousands of times more water than all known surface rivers. Such can be measured, and spring thaws predicted, for the subsequent management of water levels in storage lakes.

Life patterns in the ocean and feeding beds for fish may be detected.

Drought, crop failure, famine, and overpopulation are even now causing extreme suffering and human misery in India. We cannot afford to neglect exploiting the potential which satellites have for helping to forestall similar anguish when the population of the world reaches 10 to 15 billion people.

The tools to perform such work are in simultaneous use from orbital stations—sophisticated photography, remote sensing of a wide-band of the electromagnetic spectrum, side-looking radar, etc. With such sensors and necessary data acquisition and processing facilities, manned satellites can become one of the most useful and powerful tools to enable man to manage his Earthly resources and help stave off drastic consequences of the population explosion.

So, let me follow that advice tonight and tell you briefly about some of our activities at the Marshall Space Flight center, and the progress we are making on our plans to achieve the national goal of a manned lunar landing within this decade.

I would like to take just a few moments to explain the significance of the NASA-Marshall Space Flight Center to *all* Alabamians.

The entire state benefits from Marshall Center operations.

Booming conditions in any one part of the state affect a widespread general area. Industries throughout Alabama which produce retail items have an expanded market. Money put into circulation in Huntsville flows into other Alabama cities through increased orders, through contracts for goods and services, and as our Marshall Center personnel go back to home sites throughout the state to visit and shop. State income tax payments in North Alabama are used for improvement of services in all areas. Annually, $1,500,000 is withheld from the paychecks of our civil service workers for income tax. And they pay about $3 million annually in state and county sales tax.

We ended the year at the Marshall Center with about 7,000 civil service employees in the Huntsville area with an annual payroll between 75 and 80 million dollars. Estimates of the total employment in Huntsville created by NASA space activities run as high as 25,000, with an annual payroll exceeding 200 million dollars. Besides civil service, this includes all prime and support contractor employees, and construction employees who are working on space-related jobs.

Our employees commute to work from a roughly triangular area bounded by Birmingham, Florence, and Chattanooga. In a recent survey that included only the 7,000 civil service workers, we found that only about 64 percent listed a Huntsville home address and about 71 percent gave Madison County, 412 in Marshall County, and 363 in Limestone County.

Naturally, the greatest growth is in Huntsville, but the economic benefits are not confined to Huntsville and Madison County alone.

Marshall Center contracts and purchases are not confined to any one portion of the state. The total of more than $623 million dollars is scattered from Florence to Dothan and Mobile. The space industry in Alabama is of statewide benefit. Hundreds of purchase orders go out monthly to all parts of Alabama cities because of continuously running NASA and Marshall Center contracts. And these figures do *not* include Army activities at Redstone Arsenal, where more than 10,000 civilians take home a total payroll of more than $100 million.

While there is a very real affect upon Alabama of the national space exploration program, NASA's total outlay of money is less than one percent of the Gross National Product. We are most fortunate to live in an economy that is growing. Twelve years ago, when the Gross National Product was $362 billion, the expenditures for national defense, atomic energy, and NASA's predecessor, the National Advisory Committee for Aeronautics, totaled $47.1 billion. This was 13 percent of the Gross National Product.

In the current fiscal year, the projected total was increased to $62.26 billion, but since the Gross National Product has climbed to $700 billion, the percentage of national expenditures for space, atomic energy, and defense is only 8.9 percent.

Apollo-Saturn 201 launch vehicle was launched this past February 26 and met all test objectives. It is the Saturn IB, the first version of Saturn that will carry astronauts to Earth orbital missions. This launch will be used by NASA for manned space flights beginning about the time the current Project Gemini is completed near the end of this year. If its first flight can be used as a guide, then the Saturn IB is going to be a particularly good vehicle, performance-wise.

The next Saturn IB launch will place the second stage in orbit with a large reservoir of liquid hydrogen fuel remaining in its tank. Very little is known about the action of liquid hydrogen in a weightless environment. We want to see if we can gain and maintain control of this supercold fuel while the S-IVB stage is coasting in orbit. This is something that we will have to be certain about before the lunar exploration can get under way. The stage will be heavily instrumented, including two television cameras to provide views of the tank interior.

The S-IC is the first stage of the Saturn V Moon rocket, assembled at the Marshall center. We hope to fly the first Saturn V—without men aboard—early next year. The first stage has five engines, which produce a total of 7.5 million pounds of thrust. This is the energy equivalent of 110 million kilowatts—or about ten times the peak capacity of all the steam and generating facilities of the TVA [Tennessee Valley Authority].

We built four of the stages here in Alabama, two ground test stages and the first two flight stages. The remaining stages are being produced by the Boeing Company in our Michoud Assembly Facility in New Orleans.

We have completed the testing program on the first stage and it is being prepared for shipment to the Kennedy Space Center in Florida. The second flight stage will be going into the stand soon for its pre-flight static test program.

The S-II is the Saturn V's second stage. It is being developed by North American Aviation at Seal Beach, California. The first flight version of this stage is at Seal Beach and is being prepared for static testing that will take place at Marshall Center's test facility at Hancock County, Mississippi. The stage is expected to be ready for shipment to Mississippi this spring or early summer. All flight versions of this second stage will be tested at the Mississippi test site.

The S-IVB is the Saturn V's third stage. It is contracted to the Douglas Aircraft Company. It is also used as the second stage of the Saturn IB and is similar to the stage that will be in the limelight on the next Saturn IB orbital mission. In its Saturn V version, the stage will have a restart capability for its 200,000-pound-thrust engine. This ability to restart is the reason that we must know about the action of liquid hydrogen in space and be able to control it. During the lunar flight, the Apollo spaceship will remained linked physically to the third stage during the initial Earth orbital period. After checkout, the stage will be restarted to place the three Apollo astronauts into a course that will take them to the vicinity of the Moon. It is imperative that the stage perform as desired at this critical point, for the safety of the astronauts, as well as the success of the lunar mission.

The IU, Instrument Unit, the brains of our Saturn V, is being assembled right here in Huntsville by the International Business Machines Corporation. The Saturn IU is almost identical in both the Saturn IB and Saturn V versions. It contains all the guidance, control, sequencing and telemetry equipment for the launch vehicle.

We are not too many months away from the first delivery of all flight stages for the Saturn V. Technicians at the Kennedy Space Center will assemble the vehicle and the first Saturn V Moon rocket will be launched, in an unmanned version, early next year.

Naturally, we have encountered problems as we have moved along during the years of development since a trip to the Moon was adopted as a national goal in 1961. But, we have not, and will not, encounter a problem that cannot be overcome.

The Marshall Center, and Alabama, will meet its commitment in space. When the first American astronauts land on the Moon, it will be an epic achievement. It is difficult for us to visualize beforehand just how significant this event will be. For the first time since his creation, man will leave the Earth which has always been his home, and set foot on another heavenly body. Although this will be a magnificent achievement, even more wondrous events lie in the future. To someone who will live thousands of years from now, looking backwards over the chronology of space events, man's first trip to the Moon will pale in significance when compared with his first journey out of our solar system to the nearest star. The golden age of space exploration still lies far ahead of us.

*** END ***

— Chapter 36 —

The Overseas Press Club of America was founded in 1939 in New York City by a group of foreign correspondents. In April 1969 a contingent of about 50 members flew to Huntsville and requested a tour of Marshall Space Flight Center. Von Braun welcomed them with the following remarks.[89]

I AM DELIGHTED TO WELCOME YOU TO THE MARSHALL CENTER ON YOUR VISIT TO Huntsville, Redstone Arsenal, and the Tennessee Valley.

A lot of history has been made during the past two decades in the places which you will visit, and a great deal of activities here have influences the field of international relations.

From the national defense standpoint, Redstone Arsenal has been the center of the U.S. Army's rocket and missile programs since 1950. It is the nerve center for the Army's research, development and industrial procurement of its entire family of rockets and missiles. It contains the national inventory control point for their worldwide logistic support. And the Army school here is the center for training U.S. and NATO troops in these systems.

The Army's capability in rocketry was called upon for the entry of the United States into space. You may recall some of the famous names: *Explorer I*, our first satellite; *Pioneer IV*, first probe to the lunar region; and Mercury-Redstone, that placed the first American astronauts into space.

This is also the birthplace of the Saturn rockets used in Project Apollo, the nation's manned lunar landing program. The first boosters were built and test fired here before the job was turned over to industry.

Many of the people who now make up the Marshall Center were engaged in these Army projects before transferring to NASA in 1960.

We are proud of our past, with the Army and with NASA, but we are more interested in the future. Right now we are looking forward to the Apollo mission which is scheduled to begin with the launch of our next Saturn V at 11:49 a.m. EST, May 18, from the Kennedy Space Center. The Saturn V is on the launch pad now. The Flight Readiness Test was completed April 5. The countdown Demonstration Test, the last major hurdle before launch, began yesterday (April 27). The first part of this test will be completed May 2. The crew, consisting of Tom Stafford, John Young, and Gene Cernan, will take part in the test on May 3.

The eight-day *Apollo 10* mission will begin on May 18, with the first opening of the launch opportunity. As you know, five possible landing sites have been selected in Project Apollo from the photographs of the Moon made by Mariner, Surveyor, and the Lunar Orbiter unmanned spacecraft. Launch time is planned to permit inspection of these sites under the best possible lighting conditions. The *Apollo 10* flight will not attempt a lunar landing. With the exception of actual landing, however, the flight will be largely the same as a lunar landing mission. The main purpose of this trip is to gain additional experience in operation of the combined Saturn/Apollo systems.

When the *Apollo 10* spacecraft gets to the Moon, it will enter an orbit of 60 x 170 nautical miles. After two revolutions the orbit will be circularized at 60 nautical miles. On the fourth day of the mission, while Young continues to circle the Moon in the command module, Stafford and Cernan will enter the lunar module, separate the craft, and approach twice within 50,000 feet of one of the preselected sites. Without landing, they will rejoin Young and reenter the command module.

It will be impossible for the astronauts to land on the Moon in May and make a subsequent liftoff from the Moon because the amount of propellant in the ascent stage of the lunar module will be limited on this flight.

After the crew has inspected the Apollo landing site, they will make eleven or more revolutions of the Moon, during which they will make landmark sightings, take photographs, and transmit live television views of the lunar surface, the Earth as seen from the Moon, and their own activities inside the command module. If all goes well with *Apollo 10*, the first lunar landing may be attempted on the *Apollo 11* mission. So, before the end of July, two Americans should leave their spacecraft and step out on the surface of the Moon—if they are not held up by having to go through Russian customs.

The astronauts will remain on the Moon's surface less than one day and the time spent outside of their spacecraft will be limited to two hours and forty minutes. They will make photographs and collect samples of lunar material which will be studied later by some 130 scientists from nine countries. And they will position three lightweight experiments on the surface which should, despite their simplicity, provide valuable information on lunar seismic activity and the solar wind, and reflect laser beams from Earth long after the astronauts return home. The first manned lunar landing in Project Apollo will not mark the end of a program, but the beginning of manned lunar exploration. We hope to make many more trips to the Moon. The scientific returns from the first journey will be relatively small, for its chief objective, like Lindbergh's crossing of the Atlantic, is to prove that the trip is possible.

Using the Saturn/Apollo hardware, with minor modifications, the stay time on the Moon can be lengthened, payload weight can be increased, more scientific equipment and shelter can be provided for lunar explorers, and the transverses of the Moon's surface by scientist-astronauts can be extended with roving vehicles.

NASA is also deeply interested in continuing the unmanned exploration of the planets. And space activities in Earth orbit will be greatly increased during the next decade, as research is continued and applications are expanded. Communications via satellite is now routine by the news media. The accuracy of long-range weather forecasts will be increased.

During the next decade we will see the beginning of large manned space stations in Earth orbit. From these platforms in space, we are getting a new look at an old planet. One of the most promising areas of space exploration is the observation of Earth's resources through cameras and a wide variety of remote sensing devices. The photographs made by multispectral cameras aboard the *Apollo 9* spacecraft have created a great deal of excitement over this potential. Remote sensing devices can help reduce disease in crops and forests; locate underground water, oil, and minerals; help find fish in the oceans; improve maps; detect the dumping of manufacturing wastes into inland streams; measure ice and snow for better water management of lakes and reservoirs; measure soil fertility and salinity; and predict crop yields on a worldwide basis to match supply with demand and help prevent famine.

The potential benefits of space exploration are both exciting and unlimited. I am confident that the achievements of the second decade of the space age will surpass the accomplishments of the past ten years. The exploration of the entire solar system is a major goal of mankind which will continue to be pursued indefinitely.

*** END ***

– Chapter 37 –

On July 20, 1969, Neil Armstrong and Buzz Aldrin completed preparations to land on the Moon in the Lunar Module, *Eagle*. The two astronauts guided the Eagle into elliptical orbit. Armstrong throttled the engine at 4:05 p.m. to slow its descent. Aldrin called out altitude readings: *"750 feet, coming down at 23° ... 75 feet, things looking good ... lights on ... picking up some dust ... 30 feet, 2½ down ... faint shadow ... four forward ... drifting to the right a little ... contact light ... O.K. Engine stop."* The world heard Armstrong's message: *"Houston. Tranquility Base here. Eagle has landed."*[90]

Lunar Landing Celebration Dinner at Carriage Inn
Huntsville, July 26, 1969 [91]

I WANT TO GIVE SPECIAL THANKS TO GENERAL PHILLIPS FOR VISITING WITH US. HE not only introduced management principles, he was a shining example of dedication. His decisive role in the program is known throughout the land. And I know the United States Air Force is getting a really top-notch man. Thank you, General Phillips, and our sincere best wishes to your future.

We are here tonight to commemorate the success of the *Apollo 11*. But more than that, we are here to acknowledge a decade of hard work in preparing for the historic mission just ended, a decade of cooperation between a NASA installation and a southern town that culminated in the development of the world's largest and finest space vehicle—the Saturn V.

First I want you to know—as the men and women who have been in positions of leadership both at the Marshall Center and in the mainstream of Huntsville life—that I have appreciated the opportunity to work with each one of you and with all the aerospace workers and citizens of this North Alabama community.

Huntsville, Alabama, and the Marshall Space Flight center will live in history alongside Astronauts Neil Armstrong, Michael Collins, and Edwin Aldrin, members of the *Apollo 11* crew, and the now-famous Tranquility Base, man's first encampment in space.

The lunar landing flight and return was really and truly a beautiful thing. The names, *Columbia* and *Eagle* could not have been more appropriately chosen. And I know that I shall always remember the wonderful simplicity of the statements by the astronauts which, at the same time, had the ring of immortality. Who can forget the thrill of hearing the voices from the Moon saying, "The Eagle has wings." And then the tenseness of the actual landing on the Moon followed by the calm voice, "Earth: The Eagle has landed."

And then Neil Armstrong said all that was needed as he took the first step upon the Moon: "That's one small step for man and one giant leap for mankind."

On the day of recovery, the President stood vigil on the Carrier *Hornet* as the *Columbia* returned to Earth. After the crew was safely aboard the carrier, inside the small mobile chamber, President Nixon, as you remember, came down to speak to them. That scene will always live vividly in my memory. "I'm the luckiest man in the world," the President told the three crew members standing at the window of the small chamber. I think he spoke for all Americans.

And then he spoke for all the world: "This is the greatest week in the history of the world since creation," he said. "As a result of what has happened this week the world is bigger; the world has never been closer together. We thank you for that."

Our share of the work in Huntsville in the Apollo mission came largely in the years prior to this week's splashdown. However, it was vital to the eventual success. We worked together, and together we accomplished our part of the mission. The Moon is now accessible. And someday, because of the beginning that we have made here, the planets and stars may belong to mankind. This reach toward the heavens, toward the stars, can eventually loose the human race from the confines of the Earth and maybe even the solar system, and give it immortality in the immense and never-ending reaches of space.

These are high-sounding words. But, I think it can hardly be denied that as the space program extends the reach of man, it also expands the mind of man. We are extending this God-given brain and these God-given hands to their outermost limits and in so doing, there will be benefits and harvests to be reaped by all persons on Earth.

What do we have now that Neil Armstrong and Edwin Aldrin have stepped down upon the Moon's surface? I think that it has caused a new element of thinking to sweep across the face of the Earth. For the first time, life has left its planetary cradle and the ultimate destiny of man is no longer confined. When the Mayflower landed on American shores the pilgrims did not envision the nation that would eventually evolve. Neither can we truly say what will eventually spring from the footprints around Tranquility Base.

But I do hope that history will record that we are aware of what we have done, that we know of the enormous implications of the lunar journey, and that while we take pride in this achievement in which Huntsville has assisted, we share it in genuine brotherhood with all nations and with all peoples.

What we have now is the beginning of an exploration of the Moon that can and should continue for many more decades and eventually culminate in the productive use of the Moon—its resources and its position as a base for further steps into space.

We have begun the trip that can pave the way for missions to Mars and other planets because we know that our astronauts can live and function in deep space.

We have, in *Apollo 11*, a mission that can lead to laboratories and space stations in Earth orbit where there is such a vast potential for improving man's lot that it staggers the imagination. And Apollo is the journey that will help lead the way to new and better types of space transportation systems, that are lower in cost and will improve to the point that flying to and from orbit becomes almost as commonplace as flying across oceans in jetliners.

Those are the challenges that confront us now. And if we are to face astronauts Armstrong, Aldrin and Collins with a vow that the risks they took were worthwhile, then we must continue along those lines, and keep working toward goals for which they so bravely have made the beginning. I think it will be hard to face them with any other word than the assurance that what they have accomplished in space will be followed up with a space program second to none.

As President Nixon has said, we can "stand taller" now because of those footprints on the Moon. And we certainly see the truth in the President's words, "that every man achieves his own greatness by reaching out beyond himself, and so it is with nations." This nation said it was going to the Moon and then believed itself. And as long as America will reach out beyond the commonplace, it will make dreams come true. It will bring to future ages the bright prospects that are promised through space exploration; and by this I mean all people of all nations. The great tasks that lie before us call for a new era of international understanding and cooperation.

Knowing this, I think we can cheer tonight for a space mission that brings a new dimension of hope, a new Age of Discovery, and a new attainment for all mankind. Personally, if I were never allowed to participate further in the space program, I feel that the experiences of this past decade have been worth every moment; I would have lived my life no other way.

However, I am already looking forward toward an exciting future for the world, for the United States, and for me and for you in Huntsville, Alabama.

Thank you.

*** END ***

– CHAPTER 38 –

"I left NASA ... looking for a suitable spot in industry; I wanted to be helpful in expanding the tremendous potential of communication satellites, in particular their use for audio-visual education. It was for this reason that I joined Fairchild Industries, which has long been a leader in this field ..."

Von Braun recognized the "tremendous potential" of communications from satellite orbit many years ago, and he used every opportunity to promote the idea of using orbiting spacecraft for that purpose. Others joined his campaign, most notably Arthur C. Clarke, who is credited with the original idea of putting a communications satellite into an equatorial geosynchronous orbit, where it would remain above the same point of the equator indefinitely.[92]

NASA developed a family of Application Technology Satellites (ATS)[93], to serve the specific needs of groups of people who, by geographic or other circumstances, encounter difficulties in catching up, or keeping pace, with the development of civilization in other countries. NASA selected Fairchild Industries as the contractor to help develop and build the ATS satellites.

The ATS was launched into geosynchronous orbit in 1974. After some testing and subsequent use as a television relay station over the Western Hemisphere, including Appalachia and Alaska, it was moved within its geostationary orbit and placed over the Indian subcontinent.

In a speech to the Civitan International Convention, von Braun relates his dreams of expanding access to education to remote regions of the world.

CIVITAN INTERNATIONAL CONVENTION
ATLANTA, GEORGIA JUNE 30, 1970

PRESIDENT LEDBETTER, MRS. LEDBETTER, DISTINGUISHED GUESTS, MEMBERS OF Civitan, ladies and gentlemen:

It is indeed a pleasure for me to address an organization such as yours, which is dedicated to service without thought of reward other than the satisfaction of answering the needs of others. It is this kind of selfless service that offers the world's best hope of coping with the many problems of today. If I may paraphrase a recent remark of my boss in Washington—the NASA Administrator, Dr. Thomas Paine: "You are the kind of people who turn out and turn to, you don't turn in and turn on."

I'd like for the next few minutes to look at some accomplishments that we have made in space in the last decade and then proceed to tell you something about the promising near future developments which should help us make a better world and a better place in which to live. Let us begin with the dominating event of the past decade—the first landing on the Moon.

It is already clear today that the scientific results of both *Apollo 11* and *Apollo 12* have far exceeded even the fondest expectations that the geologists and cosmologists had when they loaded our astronauts, in their little LM [Lunar Module], down with all sorts of research gear for their walks on the lunar surface. One of them put it very aptly when he said the rocks that were brought back from the Moon have turned out to be the "Rosetta stone of the universe." The Rosetta stone, for those of you who may not remember when you learned that in history class, was a famous stone relic found near the city of Rosetta in Egypt that provided the key for the deciphering of the hieroglyphics inside the Egyptian pyramids. It thus opened a whole new world to the scientists studying the history of the Near East.

Nevertheless, I believe the ultimate significance of the lunar landing has yet to be realized. Just imagine once more what a fantastic thing it really was. Despite all the science fiction you may have read, here, in reality, a few intelligent beings from one planet actually traveling through the vastness of outer space, landing safely on another heavenly body and returning back to their own. History will, without any doubt, rank this feat of the crews of *Apollo 11* and *Apollo 12* among the great exploratory voyages, along with the travels of Marco Polo, the exploration of Columbus, or the first visit to the South Pole by Roald Amundsen.

Along with the historic event of men walking on the Moon, there came another accomplishment, which I think may have been of equal importance. I am talking about an accomplishment also made possible through the space activities of the last decade: in every corner of the Earth, men watched the living event of the lunar landing on the face of a television screen. I think it is worth remembering that only the worldwide telecommunication satellite network working together with the telecommunications equipment carried aboard the Apollo spacecraft and the receiving equipment in NASA's globe-spanning ground network made this viewing possible. The true significance of the lunar landing for man, then, is that Apollo is symbolic of the United States' peaceful efforts to bring men of all nations a little closer together in this world. The people around the globe who watched the astronauts set foot on the Moon thought of them not just as Americans up there, but as human beings—representative of all men. And let us not forget that the astronauts themselves left behind them a plaque on the Moon which read: "We came in peace for all mankind." For a brief moment, somewhere between 500 million and one billion people, in all parts of this world, paused to feel a unity in their common identity. I ask you, what other conceivable event at this stage in man's development could have brought about such a moment of human sharing?

You all know that the successful first landing of *Apollo 11* was soon followed by *Apollo 12*, which not only landed on the lunar surface but made a pinpoint landing near another spacecraft that NASA had launched to the Moon a few years earlier. And then came *Apollo 13*—a long three days—the world was glued again to its television sets, but this time wondering whether they would safely make it back.

The flight of *Apollo 13* raised the question around the world: "Is it worth the risk?" Should we really risk the lives of three brave men to conduct a voyage like this? I think there can be only one answer to that question. Exploration has always involved risks. The early settlers who came to this country with a one-way ticket, not knowing what was ahead of them, and getting into their covered wagons to cross the prairies and the deserts, and to brave the Indians—they knew what the risk was, and they faced it gladly. It was the same in the early days of aviation—without the risks our early aviation pioneers were so ready to take, there would be no worldwide air traffic today that enables you to travel within a few hours anywhere on the globe. And there were risks involved in conquering the jungles and the oceans and the poles. The crew of *Apollo 13*, when asked upon their return, said they would be most ready to go again, if they would only be given a chance. Of course, we have so many astronauts standing in line now who *all* want to get their chance to fly, that the members of the previous Apollo crews will have to wait quite a while before they could fly again, and some of them may reach their age limit and wind up behind some desk, administering part of the space program so that others can have the fun.

There is another question that is heatedly debated today: "Is the space program worth the cost?" People point to the competing social needs and ask whether there aren't more important things to do in this world than to fly to the Moon or to conquer space. I would like to give you a few figures. In 1958 the Gross National Product of the United States of America was $440 billion. In 1969 the same GNP was $900 billion. So it grew by $460 billion, or more than doubled, during the first 10 years of the Space Age. I do not believe that it was purely coincidental that this phenomenal growth in our GNP occurred during the same time span we set out to get our space program rolling. I think you will all agree that the growth of the GNP—the productivity of this country, its ability to produce goods and sell goods—is really the key to all our social advances. We cannot hope to solve the problems of urban decay or to clean up rivers or to unclog the highways if we didn't have the funds to do it. These funds come from taxes which can be raised only if we have a strong tax base supported by a strong and vibrant industry. The GNP and its growth was the key to all the progress we have seen in the social field in the last ten years. And why has the space program so greatly contributed to the growth of the GNP? Because it compressed several decades of technological development into just one. This has given our industrial capability a tremendous shot in the arm. The space program has made America richer and *not* poorer.

Let me substantiate this claim with just two figures. The computer industry is now an $8 billion a year business. Today, virtually every on-line direct access commercial computer anywhere in the world is U.S. made, and I might add "courtesy of the space program." The computer industry enjoyed a 1400% growth during the first decade of the Space Age. Another example is the aviation industry. It is actually a lot larger than the computer industry. The aviation industry sold $27 billion worth of airplanes and associated equipment last year, and of course, it is the main industry supporting the space program. For that reason it is also the most direct beneficiary of the new technology and know-how emerging from the space program. I think the results are evident. Wherever you travel in the Free World today, you find an abundance of United States-made airplanes. And yet it is not easy to sell airplanes or computers all over the world, in spite of the fact wages in America are up to 10 times higher than in some of our competitors' countries. So if we are nevertheless able to dominate the export market for this kind of equipment, our product must be awfully good. It is an established fact that our large positive international balance of trade today has been possible predominately through the exports of our aerospace and computer industries. It is these two industries that still enable you to freely travel overseas, that enable our ladies to buy Italian shoes and us men to drink Scotch whiskey.

But the question still remains, should we not pay more attention to our domestic problems, and here my emphatic answer is yes, we should. We should indeed pay much more attention to our domestic problems, but this does not mean for one moment that we should abandon space or even weaken our resolve to explore and utilize this fascinating and fertile new arena of man's activities. Let's look back at the early dawn of man's history, the time when man still lived in caves. Some of those cavemen must have looked out into the darkness from their caves and wondered what was beyond those blue hills on the horizon. But the problem was that the caves were leaking—puddles of water had accumulated at the bottom of the cave and the sanitary conditions were deplorable. There were those in the cave who said that the exploration-minded fellows should stay right here and first take care of the cave's domestic problems—fix the leaks and get rid of the sewage. But then there were some who just couldn't suppress their curiosity about those shiny blue hills in the distance. I think we would probably still be sitting in caves today if these wild blue yonder men had not prevailed. Maybe they decided that they could fix their caves *and* explore the hills, or maybe they were even smart and advanced enough to persuade their wives to fix the leaks so they themselves could do the exploring.

These kinds of domestic problems have always been with us. I think we can be certain that the little harbor town of Lagos in Spain, from whence Columbus departed, had a pretty serious pollution problem too. Should Columbus have been persuaded to first clean up the harbor before he set sail?

Today many people ask: "If we can go to the Moon why can't we—" and then comes a list of all our national domestic ills. Let's analyze this question for a moment. While we certainly do not completely understand all scientific problems involved in ecology or all engineering problems, but solving a domestic issue is not a technical problem, but mostly one of wise legislation and law enforcement. You see, going to the Moon was a clear-cut technological and scientific challenge, and it was a formidable managerial challenge, too. But since no people live between the Earth and the Moon there was very little involved in what you might call a conflict of interest between human beings. We did not step on anyone's toes.

But now take a typical domestic problem; take the case of city smog. The scientific answer to the smog problem is very simple. Stop driving! Well, clearly people must continue to drive in order to get to their jobs. In today's world millions would starve to death if they were deprived of their automobile. Or take airport noise; here again, the scientific answer is very simple: ground the airplanes and there will be no airport noise. But without air travel, our tightly-knit economy would take a nose dive and hundreds of thousands would soon lose their jobs. Or river pollution; stop dumping unprocessed wastes into our rivers! But the problem is that our industry operates in a highly competitive environment. Clearly, unless waste management legislation can be developed which applies uniformly to all factories manufacturing the same products, no matter where they are located in the United States, you can easily legislate a factory right out of existence. If a pulp mill in one state is told it must install a very expensive waste treatment plant if it wants to continue its operation, while a competing mill in another state goes scot-free, the former plant will soon have to close down. The result of such uneven ground rules will be local unemployment. There have already been scattered incidents where laid-off workers were seen with picket signs: "I lost my job. Are fish more important than people."

This kind of backlash could seriously hurt our badly needed environmental protection program before it even got off the ground. So, what we need is wise, well-researched, unemotional, and uniform pollution legislation. But the scientific and technological base on which such legislation must be founded, is usually there.

While we in NASA thus cannot be of much help with this type of problem, there exists one area where we have built up substantial competence and which may indeed come in handy to help fight the environmental problems. Here I am referring to NASA's expertise in the hostile environment of outer space. In space there is zero gravity—pretty hostile to those who are not used to it. There is a vacuum which would kill you outright if you are not protected by a pressurized space cabin or a space suit. There is deadly radiation out there from which you must be adequately shielded. There are extreme temperatures, hot and cold, depending on whether you are in sunshine or shadow. Then there is the entire aeronautical flight regime—coming back into the atmosphere at Mach number 20 until you finally deploy your parachute at subsonic speed—undoubtedly a very hostile environment and utterly unhealthy for anyone exposing himself to it without adequate protection. Of course, we protect our astronauts against all the dangers resulting from this hostile environment and keep them safe and comfortable with the help of some special closed-cycle support equipment.

NASA can indeed extend the studies that lead to this protective equipment to other types of hostile environments for man. Hostile environments such as crowding, malnutrition, air, water, noise, pollution, but also violence, insecurity, fear, economic dislocation, resource depletion and even natural disasters such as quakes, hurricanes and tornadoes. And from such studies we could likewise derive suitable protective measures against some of these factors. The approach common to studies on "man vs. the hostile environment in space" and to "man vs. the hostile environment on Earth" is what we call the systems approach.

But instead of getting too deep into methodology, let me give you just one example of how the space program can make very direct and meaningful contributions to the solutions of Earth-related problems. My example involves an educational television system that the United States has agreed to jointly bring into being for and with the help of India. In November 1969, the NASA Administrator and Dr. Vikram Sarabhai, the Chief of the Science Programs of the Indian government, signed a treaty under which NASA will establish a satellite in synchronous orbit over the Indian Ocean. Now a satellite in synchronous orbit is a satellite with a period of revolution of 24 hours, and since it orbits in the plane of the equator traveling from west to east, it will stand still over one point of the equator. This is simply because it revolves through its orbit just as fast as the Earth rotates underneath. So from a point anywhere in India, that satellite will be at one fixed point in the sky, day and night, rain or shine. Lest you think that the establishment of such a synchronous satellite is a great novelty, let me hurry to add that we have already something like 20 satellites of this kind in various points of an equatorial synchronous orbit. They are used for transoceanic communications purposes. So this is old hat as far as the space technology is concerned.

This specific India satellite—we call it ATS-F—will be sent up in 1973. It will have a receiver and a transmitter and it will be capable of receiving and transmitting simultaneously two educational television programs. These programs will be beamed up to the satellite from an Indian ground station near Ahmedabad, which is somewhere north of Bombay. The satellite receiver picks up these programs, the satellite transmitter then sends them back with the help of a big 30-foot dish antenna whose range covers the entire Indian subcontinent. Now, in India proper, 5,000 television receivers will be furnished by the Indian government and distributed to 5,000 villages. These receivers will have a special input end for the ultra-high frequency used by the satellite transmitter, but otherwise they will look exactly like a normal television set. This type of set will be located in the village marketplace or, if available, in a community building. The Ahmedabad transmitter station and the 5,000 receivers will be manufactured by Indian electronic manufacturers, thus providing a shot in the arm for India's own electronics industry. One almost amusing aspect stems from the fact that the majority of these 5,000 villages in India do not have rural electricity. So in order to provide the puny 200 watts to drive the television receiver, a stationary bicycle will be delivered with it. An Indian villager pedaling away will thus produce the power to extract the educational messages from a station in the sky, over 25,000 miles away.

The programs that will be viewed on these community receivers will include several hours of reading, writing, and arithmetic for the kids in the morning. There will be farm instructions in the afternoon where farmers will be advised what seed goods to acquire for their particular soil, what fertilizer to use, and how to irrigate more effectively. Somewhere in between may be a session called family planning that will deal with such things as how to use the "pill." A local educational TV program of this kind in an area around New Delhi has turned out to be immensely popular, and the idea to extend this kind of service with the help of our synchronous satellite has found tremendous response throughout the country. The reason for this favorable response is that education is something very dear to the hearts of the average farmer in India. They have a saying that in India the word education is as holy as the cow.

Now, I would like to point out that the acquisition of those 5,000 village receivers working with our satellite in 1973 is again only an experiment. The treaty provides that our satellite service will be rendered for one year. But it is open-ended and can be extended for more years if both sides like it. The Indian government really considers this experiment merely as a forerunner for a much more ambitious program. India, after all, has not 5,000 villages but, believe it or not, 539,000 villages, and it has not just 5 million people which the 5,000 receivers may be able to service, but 540 million people. The great majority of these half-a-billion people live in villages. And most of these villages are without access by an all-weather road, let alone rural electricity or local television. If the interim "experiment" is successful, the Indian government plans to go all-out and make this satellite educational television system the mainstay of the educational system in India. Technologically underdeveloped India will thus be bypassing the telephone, the country road, the highway, the railroad, the airplane and even rural electricity, and it will go directly to the most advanced communications system ever devised by man—the voice from the sky—to get the knowledge spread among its people. India found that this was the only way to reach its huge farm population. Any idea of developing a system as we have in the United States with local television stations had to be discarded because the individual Indian farmer just could not afford to buy a private TV set, or the detergents whose advertisements would support the station. Only with government-furnished community sets and a central program station would it be possible to reach the hundreds of millions of farmers.

*** END ***

— CHAPTER 39 —

RELIGIOUS IMPLICATIONS OF SPACE EXPLORATION: A PERSONAL VIEW [94]
BELMONT ABBEY COLLEGE BELMONT, NORTH CAROLINA, 22 NOVEMBER 1971

FOUR APOLLO CREWS HAVE NOW SUCCESSFULLY RETURNED FROM THE MOON. A fifth one, while not succeeding in getting there, returned hale and hearty to Earth. Two more Apollo flights are scheduled for next year.

Why are we flying to the Moon? What is our purpose? What is the essential justification of the exploration of space? The answer, I'm convinced, lies rooted not in whimsy, but in the nature of man. Let me give you my personal view.

Man as a biological species is a rather anomalous animal. Whereas all other animals establish a place for themselves in nature's highly cooperative and competitive ecological system, man has established his place in nature by altering his natural environment through such actions as practicing agriculture rather than eating the natural fruits of the trees and the plants, and clearing forests to build cities.

Whereas all other living beings seem to find their places in the natural order and fulfill their role in life with a kind of calm acceptance, man clearly exhibits confusion. Why the anxiety? Why the storm and stress? Man really seems to be the only living thing uncertain of his role in the universe; and in his uncertainty, he has been calling since time immemorial upon the stars and the heavens for salvation and for answers to his eternal questions: Who am I? Why am I here?

Wherever he fought, he invoked the stars for help. Wherever he loved, he invoked the Moon. And all great religions hold out eternal life and heaven as man's reward for his good deeds on Earth.

Whereas most animals follow, for their survival, certain telltale scents which are too refined for human perception, man seems to be uniquely equipped to perceive certain vibrations emanating from the celestial environment. As a result, astronomy is the oldest science, existed for thousands of years as the only science, and is today considered the queen of the sciences. Although man lacks the eye of the night owl, the scent of the fox, or the hearing of the deer, he has an uncanny ability to learn about abstruse things like the motions of the planets, the cradle-to-the-grave cycle of the stars, and the distance between the stars.

Whereas all other species seem resigned to the environments in which they have been born, man clearly [is] not. Since his early beginning he has wanted to fly, and today he does fly.

What is man's motivation? Why does he always want to explore what is outside his abode? Why is he so eager to pioneer activities beyond his natural endowments? Why is he so interested in science?

As I said before, the mainspring of science is curiosity. Since time immemorial, there have always been men and women who felt a burning desire to know what was under the rock, beyond the hills, across the oceans. This restless breed now wants to know what makes an atom work, through what process life reproduces itself, or what is the geological history of the Moon.

But also, there would not be a single great accomplishment in the history of mankind without faith. Any man who strives to accomplish something needs a degree of faith in himself. And whenever he takes on a challenge that requires more moral strength than he can muster with his own limited mental and spiritual resources, he needs faith in God.

But many people find the churches, those old ramparts of faith, badly battered by the onslaught of three hundred years of scientific skepticism. This has led many to believe that science and religion are not compatible, that "knowing" and "believing" cannot live side by side.

Nothing could be further from the truth. Science and religion are not antagonists. On the contrary, they are sisters. While science tries to learn more about the creation, religion tries to better understand the Creator. While, through science, man tries to harness the forces of nature around him, through religion he tries to harness the forces of nature within him.

Some people say that science has been unable to prove the existence of God. They admit that many of the miracles in the world around us are hard to understand, and they do not deny that the universe, as modern science sees it, is indeed a far more wondrous thing than the creation medieval man could perceive. But they still maintain that since science has provided us with so many answers, the day will soon arrive when we will be able to understand even the creation of the fundamental laws of nature without a Divine Intent . They challenge science to prove the existence of God. But, must we really light a candle to see the Sun?

Many men who are intelligent and of good faith say they cannot visualize God. Well, can a physicist visualize an electron? The electron is materially inconceivable and yet, it is so perfectly known through its effects that we use it to illuminate cities, guide our airliners through the night skies, and take the most accurate measurements. What strange rationale makes some physicists accept the inconceivable electron as real while refusing to accept the reality of God on the grounds that they cannot conceive Him? I am afraid that, although they really do not understand the electron either, they are ready to accept it because they managed to produce a rather clumsy mechanical model of it borrowed from rather limited experience in other fields, but they wouldn't know how to begin building a model of God.

For me the idea of a creation is not conceivable without invoking the necessity for God. One cannot be exposed to the law and order of the universe without concluding that there must be a Divine Intent behind it all.

I am also confident that, as we learn more and more about nature through science, we shall not only arrive at universally accepted scientific findings, but also at a universally accepted set of ethical standards for human behavior. In the world around us we can behold the obvious manifestations of the divine plan of the Creator. We can see the will of the species to live and propagate. We behold the gift of love. And we are humbled by the powerful forces at work on a galactic scale, and the purposeful orderliness of nature that endows a tiny and ungainly seed with the ability to develop into a beautiful flower. The better we understand the intricacies of the universe and all its harbors, the more reason we have found to marvel at God's creation.

Some modern evolutionists believe that the creation is the result of a random arrangement of atoms and molecules over billions of years. But when you consider the development of the human brain within a time span of less than a million years, simple statistical studies indicate that for random processes this span just is not long enough to produce the brain whose tremendous complexity we are only now beginning to understand. Or take the evolution of the eye in the animal world. What random process could possibly explain the simultaneous evolution of the eye's optical system, the nervous conductors of the optical signals from the eye to the brain, and the optical nerve center in the brain itself where the incoming light impulses are converted to an image the conscious mind can comprehend? Can all this really be explained without the notion of a Divine Intent, without a Creator?

Also, it is one thing to accept natural order as a way of life, but the minute one asks "why?" then again enter God and all His glory.

Our space ventures have been only the smallest of steps in the vast reaches of the universe and have introduced more mysteries than they has solved. Speaking for myself, I can only say that the grandeur of the cosmos serves only to confirm my belief in the certainty of the Creator. Finite man cannot begin to comprehend an omnipresent, omniscient, omnipotent and infinite God. In the final analysis, any effort to reduce God to comprehensible proportions beggars His greatness. I find it best to accept God, through faith, as an intelligent will, perfect in goodness, revealing Himself through His creation—the world in which we live.

Of course, the discoveries in astronomy, biology, physics and even psychology have shown that we have to enlarge the medieval image of God. If there is to be a mind behind the immense complexities of the multitude of phenomena which man, through the tools of science, can now observe, then it is that of a being tremendous in His power and wisdom. But we should not be dismayed by the relative insignificance of our own planet in the vast universe as modern science now sees it. For it is perfectly conceivable that such a divine being has a moral purpose which is being worked out on the stage of this insignificant planet. And, coming to our Christian faith, it is entirely sensible to believe that such a God deliberately reduced Himself to the stature of humanity in order to visit Earth in person.

In fact, I believe that there is much evidence that God, after He set the process of evolutionary development in motion, should have considered this process as highly experimental. The crust of our Earth is full of remnants of false starts, and many evolutionary strains, such as the big reptiles, were only temporarily successful and wound up in a dead alley.

One of the greatest experimental risks the Creator ever took was certainly when he endowed man with the privilege of free will. As the story of Adam and Eve's expulsion from Paradise so touchingly recounts, the built-in conflict between that newly authorized free will that gave man an almost god-like stance and man's old animalistic survival-of-the-species instinct promptly ran him into trouble, and has led to problems for mankind ever after. In fact, the cumulative effect over the centuries of millions of individuals choosing to please themselves rather than obeying their natural God-given instincts infected the whole planet and produced what theologians call "this sinful world."

In our limited human world, any responsible innovator who sets up a risky experiment will monitor it carefully and avail himself of all observational means to do this effectively. What could be more effective for God but coping Himself for a while with human life on the very terms that He had imposed on His creatures? By descending to Earth and spending life as a man amongst men, people would see Him not seated high on a heavenly throne, but in the battlefield of life.

It is perhaps not too surprising that once God had decided to monitor His free will experiment for a while by becoming a man Himself, the experience of His contacts with real people proved to be short of pure agony. Conversely, to the ruling establishment of the people He visited, His divine message which so effectively aroused His listeners, was a source of astute embarrassment, for it tore the veils off the half-truths and the self-serving interests of the high and mighty. In man's time-honored fashion, they would unleash the whole arsenal of weapons against Him: misrepresentation, slander and accusation of treason. Unrepentant and incorruptible as God must be, He would suffer the full impact of some trumped-up charge. The stage was set for the ironical situation without parallel in the history of the Earth. God decided to visit the planet Earth to see how His human creatures are doing with the free will He gave them, and they nailed Him to the cross.

When man, almost 2,000 years ago, was given the opportunity to know Jesus Christ, to know God who had decided to live for a while as man amongst fellow men, the world was turned upside down through the widespread witness of those who heard and understood Him.

The same thing can happen again today. I am not in despair about the discordant conditions of our social environment. In spite of all the temporary setbacks that humanity has suffered through the centuries, I strongly believe that God, in the same personal relationships He established through Jesus Christ, will see to it that man's path will continue upward, leading toward gradual improvement.

Jesus greatly expanded mankind's basic moral laws. His commandment, "to love thy neighbor as thyself," established the unselfish attitude that enables human beings to live peaceably together. Even more revolutionary was His commandment to "love thy enemies." Although it is all too rarely followed, it has left an indelible and unforgettable imprint on the man-to-man relationship among people everywhere on our globe.

Some people seem to have serious difficulties tying together certain Biblical passages with the reality of science, such as the story of creation given by Genesis, of the account of Joshua's poetic appeal for the Sun to stand still while the Israelites avenged themselves over their enemies. The interpretation of Biblical passages has been the subject of argument between wiser men than myself for centuries. My own views on the delicate topic are that it helps to bridge the gap between the

Bible and modern scientific thought if we remember that the Bible deals with man as well as God, and most of the people of whom the Bible speaks suffered from the same human frailties that we experience today.

In my opinion, (and let me emphasize here that I fully respect and honor different views) insistence on an inflexible type of religion, holding to a literal interpretation of every word of the Bible as ultimate truth will tragically delay reconciling some of the Biblical references to scientific interpretations. But I believe, with all my heart, that religion, like science, is evolutionary, growing and changing in the light of further revelations by God. While the Bible is the best preserved account that we have of the revelations of God's nature and love, we should recognize that particularly the early books, such as Genesis, were not written by scientific observers and witnesses, but by scribes who recorded ancient shepherd songs and tales because of their allegorical beauty.

A scientist who discovers a new bit of knowledge does not tear down his model of reality, but merely changes it to agree with a new set of observations and experiences. By doing so, he admits he has no claim on ultimate truth. His laws are simply observations of reality, which he is always ready to update as his enlightenment grows.

And so, I think, it should be with the Christian religion. While preserving the ethical, moral, and spiritual meaning of the scriptures, the Christian churches should become a little more flexible with regard to various interpretations of the Bible as an historical account. We can still love and have faith in words of the Bible as they reveal so effectively so many time honored truths.

The Christian churches cannot hope to reassume their rightful responsibility for ethical guidance with irrelevant debates concerning science versus religion. Although I know of no reference to Christ ever commenting on scientific work, I do know that He said: "Ye shall know the truth and the truth shall make you free." Thus I am certain that, were He among us today, Christ would encourage scientific research as modern man's most noble striving to comprehend and admire His Father's handiwork. I am certain that our Christian churches know full well that they must come to grips with the world of the Twentieth Century realities, which constitute the main concern of contemporary man. The fresh wind which is blowing through most of the world's Christian institutions is a sure indicator that the churches are indeed responding to the new and unprecedented demands placed on them by the Space Age. Ecumenical Council meetings in Rome and interdenominational conferences in the U.S. are strong evidence of a growing realization among Church leaders that the most important criteria for the future of Christianity and the guiding role of Christian ethics in these troubled times is the emphasis by a united front of the Christian religions on what Christians have in common rather than what sets them apart.

I am quite confident that the great majority of Church leaders know in their hearts that this united front can best be presented by a common faith of all Christians in the basic teachings of Jesus Christ. But it means learning to live with the findings of Copernicus, of Galileo, of Darwin. This front requires an emphasis on the essentials of spiritual life as identified by Jesus Christ rather than on trivia. It requires the acceptance of change and the discarding of antiquated or downright erroneous ideas no matter how painful.

In this reaching of the new millennium through faith in the words of Jesus Christ, science can be a valuable tool rather than an impediment. The universe as revealed through scientific inquiry is the living witness that God has indeed been at work. Understanding the nature of the creation provides a substantive basis for the faith by which we attempt to know the nature of the Creator.

When Frank Borman returned from his unforgettable Christmas 1968 flight around the Moon with *Apollo 8*, he was told that a Soviet Cosmonaut recently returned from a space flight had commented that he had seen neither God nor Angels on his flight. Had Borman seen God, the reporter inquired? Frank Borman replied, "No, I did not see Him either, but I saw his evidence."

*** END ***

– Chapter 40 –

In January 1971, about 12,000 secondary school principals from every state in the union, and from many foreign countries, met in Houston at the Annual National Association of Secondary School Principals convention. The theme of the meeting was an examination of critical issues confronting the educators as they "faced educating youth for living in tomorrow's world."[95]

National Association of Secondary School Principals

THANK YOU FOR GIVING ME THE OPPORTUNITY TO ADDRESS THIS GENERAL SESSION of the National Association of Secondary School Principals (NASSP)—representative as it is of professionals who have over the years shown a clear understanding of America's space program and its meaning to the Nation.

You have recognized the space program in its deeper philosophy and importance as something far more than simply putting men on the Moon. In my contacts with members of your profession, I have found a depth of perception that is heartening—a realization that the space program is an extraordinary, an unprecedented, search for new and fundamental knowledge. It is a search into the essential nature of existence, and of man's place in it.

How different a reaction from the time, just a few hundred years ago, when the eminent professor of Philosophy at the University of Padua refused even to look through Galileo's telescope at the satellites of Jupiter. That learned man knew, quite positively, that Jupiter could not have satellites, and what Galileo reported must then, of course, be witchcraft or magic.

You have shown your understanding and your open-hearted cooperation in many ways. We of the National Aeronautics and Space Administration are grateful that so many of you have welcomed the NASA Spacemobile Program to your school assemblies and classrooms. We deeply appreciate the support you have given to our Youth Science Congresses, and the various other projects and programs of our Educational Programs Division over the years.

But particularly, we recognize the importance of your efforts in updating the various curricula by weaving in new knowledge from space into the teaching of the sciences, industrial education, and the social studies. I am sure that most of you are aware of the very significant contribution in this area, the *Aerospace Curriculum Resource Guide*, which was initiated by your Executive Secretary, Dr. Owen Kiernan, when he was Commissioner of Education for the Commonwealth of Massachusetts.

I hope on this occasion to give you an idea of the directions of NASA's basic research, some of the understandings we have gained, and something of the magnitude of the new questions evolving from old answers. I also plan to tell you a little about the space program's applied research, and its important practical contributions to our living here and now and in the years ahead. And I will conclude with a request to you as educators, which I hope you will give your thoughtful consideration.

As I discuss the results of the space program's basic and applied research, keep in mind that the resulting new knowledge also represents the technologies, the inventions, the sophisticated new instruments and materials which the scientists, engineers, and technicians have developed to accomplish the target researches.

The indexes and bibliographies of space-related research list 60,000 to 70,000 studies each year!

As for space-related basic research: our scientists, astronomers, and cosmologists are beginning to gain insights into the nature of such fundamentals as infinity and eternity. In the life sciences, our scientists are attempting to duplicate in the laboratory the processes by which the animate was synthesized from the inanimate hundreds of millions of years ago.

Many of us here today may live to experience a revolution in man's thought as radical and as significant as those precipitated by Darwin, Newton and Copernicus.

The scientific questions being probed by NASA are related to the very nature, maybe the essence, of things. In the heavens now, making observations at wavelengths usually blanketed out by the Earth's atmosphere are:

1. The Orbiting Astronomical Observatory with its 13 telescopes.
2. The Radio Astronomy Explorer with antennas which extend 1,500 feet from tip to tip.
3. The Small Astronomy Satellite, recently launched by Italy from the Indian Ocean.

Plans call for the launching during the 1970s and the 1980s of:

Skylab: a multipurpose, three-man space station, in 1972, to examine the Sun, the stars, the Earth, and to test the ability of man to live in space for a month or more.

Viking: an unmanned Mariner Spacecraft and Lander, scheduled to probe the Martian surface in 1975.

Grand Tours: will swing by Jupiter, Saturn, Uranus, Neptune, and Pluto to take pictures and scientific reading of these planets, in the later part of the decade.

Space Stations: where scientists and engineers can work in shirt sleeve environs on scientific, engineering, and fabrication problems, forbidden to man by the Earth's gravity, atmosphere and magnetosphere.

Our observations of the heavens both from spacecraft and Earth are discovering sources of radiation, matter, and energy; sources whose nature we do not yet know or understand; sources for which we can only give such names as pulsars and quasars. Some cosmologists say these observations may give us clues to the control of never-before-dreamed-of energy—energy far greater than that made available through the mid-20th century's exploitation of the atom's nucleus.

But now let us turn from the vast reaches of the cosmos to the here and now of planet Earth.

Orbiting Earth today are ITOS satellites, operated by the National Oceanographic and Atmospheric Agency (NOAA) and developed by NASA to take cloud cover pictures of the Earth, day and night. For four years the Weather Bureau and its successors, the Environmental Science Services Administration (ESSA) and NOAA, have operated the NASA-developed TIROS spacecraft and integrated them into the Nation's processes of weather observation and forecasting. Our NASA research and development, however, continues with the Nimbus and ATS spacecraft—test vehicles for new and improved instruments. Before long, three or four weather satellites will be in geostationary orbit over the equator, 22,300 miles above the Earth, monitoring the Earth's weather continuously. They will provide more dependable weather forecasts and long-range predictions and give early warning of hurricanes, tornadoes, and typhoons; these satellites may possibly even lead to understanding the dynamics of weather control.

Also as part of this effort to predict and understand the weather are our Orbiting Solar Observatories, and other spacecraft with instruments to investigate the specific influences which the Sun has on Earth.

Also developed and under development by NASA are communication satellites. All of us here remember the first experimental TV broadcasts, and the formation of the ComSat Corporation as a direct outgrowth of NASA research. Now there is a worldwide telecommunications network to multiply twentyfold the channels of worldwide communications. Operating these NASA-developed satellites is an international, quasi-governmental consortium called Intelsat. These communications satellites are in geostationary orbit 22,300 miles above the equator; i.e., each appears to stay at the same point over the Earth's surface.

Also being developed are low-cost navigational systems using satellites as navigational aids for civilian-operated aircraft and seacraft.

But important as these developments are to life on Earth, I think perhaps there is something even more significant: the possibility that NASA may have developed the tools that will permit society to come to grips with the problems of preserving a livable, habitable environment.

Mankind today may be the beneficiary of one of those providential developments which every now and then occurs in world history. We can name a few: the discovery of fire, the invention of the wheel, the compass, the printing press, the microscope.

For the past fifty to seventy-five years, conservationists in America and Europe have called attention to the limited nature of the Earth's resources. During the past decade, environmental scientists and engineers, geographers and intellectuals began to detect evidence of a serious and perhaps irreversible deterioration of the ecological conditions that support life on Earth. There have been such jokes about conditions as: 'Anyone who takes a deep breath these days isn't very concerned about his health!' But we are long past the joking stage. Many persons have begun to believe that mankind has only limited ability to grapple with the problem; they see themselves with not even the means for awakening mankind to an awareness of the cataclysmic consequences of this environmental deterioration.

But to those with vision, there was hope—a turning point—on Christmas Eve of 1968. Astronaut Borman, as he and his crew circled the Moon, read from Genesis and followed with pictures of our planet Earth. Mankind, for the first time, saw the beautiful little blue and silver sphere on which we live shining forth against the blackness of infinity. The peoples of Earth saw for the first time how finite our habitat really is; they sensed its essential frailty. Mankind received the warning it could understand! Those of us who believe that there is an Almighty who cares for man may have abundant evidence for our faith.

Not only did the space program catalyze man's concern for preserving his environment, it is providing him with the possibilities for investigating the Earth's surface and it's subsurface. We are developing instruments that can detect blight, forest fires, sources of pollution, sources of fertility; that can locate fishing grounds, oil fields, mineral deposits; that can measure the extent of environmental degradation and can monitor its sources.

These instruments are being tested on high flying aircraft; they have been used in our Gemini and Apollo spacecraft. They will soon be orbiting the Earth on Earth Resources Technology Satellites (ERTS). When fully tested and ready to become operational, they will be turned over to a national or, perhaps a world, agency with specific responsibility for studying, monitoring and preserving the Earth's resources.

I think there is renewed hope for the future of what Buckminster Fuller calls "Spaceship Earth." And now, in concluding, I would like to discuss the relationship between what I am doing as a manager in America's space enterprise and what you are doing as managers in America's secondary education enterprise.

To help make my point, I will go back a quarter of a century into history. In the late 1940s, the lack of public concern for science and technology had become so pronounced that leading American scientists caused the National Science Foundation to be established. The support of science research became a policy of the United States. An offshoot of that policy was the support of science education. You are well aware of the results in secondary school science: special courses for bright students; increased training of teachers of science; better stocked and equipped laboratories; and science courses which broke from the cook-book teaching of college science courses to embrace the processes of scientific investigation. The press toward training scientists and engineers unfortunately became overemphasized after Sputnik. A number of us in NASA, including our Educational Programs Division, tried to resist this overemphasis. After all, Russia beat us into space with a satellite not because of our lack of know-how and technology, but because of high-level political decisions.

We see today as the Vietnam conflict subsides, a loss of public interest in science like that of 25 years ago. However, we note an additional and frightening phenomenon—it might be described as a downright antipathy toward science and technology.

I can understand this lethargy toward science education. After all, we see feature stories almost daily about engineers having to run hamburger stands or read gas meters. Indeed, many scientists and engineers are among the unemployed. I have more difficulty understanding the hostility toward science and technology, but also can understand its source.

What concerns me is that, whereas the educational pendulum swung too far toward science education in the late 1950s and 1960s, it may very well be swinging too far in the other direction.

In the late 1970s and early 1980s, society will be calling for trained engineers and scientists and a scientifically literate citizenry to make decisions and to affect the promise, which science and technology possess, for providing a better Earth environment. It is, as I indicated earlier in these talks, a promise which now is beginning to appear on the horizon. It would be tragic if mankind failed by default because it had too few scientists and engineers to take advantage of this promise.

To you as secondary principals, responsible for the secondary school education of 25 million American boys and girls, I ask that you use your influence to keep the anti-science swing of the educational pendulum from going too far. I also ask of you to effect something even more difficult—the developing of school programs that will make every student a science literate. By a science literate, I mean not a future scientist or engineer, I mean a future layman who can react intelligently, critically, and knowingly to the societal problems posed by science and technology.

A modern educational program truly reflective of the wonderful world in which we live, attuned to the times and capable of serving the splendid young people of America, is the objective of everyone in this room.

The fate of any modern educational program rests with you and those whose activities you direct.

NASA, the custodian of a large segment of the nation's effort in space, stands ready and, we hope, able to bring you the latest developments in this extensive area of research and exploration, the impact of which is just beginning to be felt.

Giving directions today to the educational process is an exciting prospect for you.

Assisting you, in a small way, is equally exciting for us.

*** END ***

— Chapter 41 —

In Osaka, Japan, NHK, radio Japan, sponsored a series of lectures by internationally recognized scholars and leaders as a public service activity. Von Braun presented the following speech in March 1971.

Fall-Out Effects of Megascience
Necessity for Huge Engineering Projects: Social Significance [96]

TECHNOLOGICALLY, THE GOAL OF LANDING A MAN ON THE MOON BY THE END OF the decade and returning him safely to Earth was a super challenge to the America of 1961. In fact, we did not realize at first how great a challenge it was.

The scientific and engineering communities of that time were generally confident that a manned lunar landing by 1970 was "technologically feasible." However, that was really a euphemism, concealing a great deal of ignorance of what, exactly, we were getting into. No one, at that point, had studied in depth the finer details of the problems.

Nevertheless, when a nation accepts a great and inspiring task, such as the lunar mission announced by President Kennedy almost ten years ago, it liberates unique talents of its people. Often these talents remain unused and dormant during times when no extraordinary demands are made either by circumstance or national decision. And this outpouring of talents and capabilities is a major effect, and benefit, of huge scientific-engineering projects having worthy and inspiring goals. It is a key factor in the operation of a technological mechanism we call "megascience."

Now, by "megascience" we do not mean just pouring a lot of money into a number of valuable but unrelated scientific investigations; we mean a very large concerted and organized attempt to achieve a technological objectives by a certain date—an objective that lies beyond existing capabilities of science and engineering over a broad spectrum of these disciplines. Such a multi-disciplinary and interdisciplinary effort, bound together by a common purpose, was Apollo.

When President Kennedy announced the lunar mission to be a national goal, the only simple and clearly understood aspects of the task were the target—that very familiar and visible face of the Moon—and the date by which it was to be completed. These proved to be among the key factors operating for eventual success of the Apollo Program. More important, they were the motivating elements that inspired both program participants and the American public.

As a target, the Moon is ideal. It is highly visible and well-known. It is concrete, not abstract. Anyone can visualize building a spacecraft, landing men on the Moon, and returning to Earth with a piece of it. To the scientist and the engineer, the Moon is located at a distance that challenges, but does not discourage, human ingenuity. They can precisely identify the requirements for a manned lunar exploration voyage.

But as we got deeper into the problems, we discovered a large knowledge gap. The need for specific data called for scientific and engineering developments that went far beyond our capabilities existing in 1961. This need for specific information started many preliminary scientific and engineering projects to obtain the data required. An extraordinary team composed of government agency, university, and industry scientists, engineers, and technical managers was formed.

The date set by the President for a successful manned lunar exploration was a vital element in the program. A lunar mission, without a publicly announced schedule to impose its discipline on every decision and action, probably could not have been accomplished. It would have crumbled away in procrastination and frustration. The public, too, would have lost interest, seeing no fulfillment of its high expectations. The schedule acted as a spur to keep the program moving. But it also helped to hold down costs that surely would have risen appreciably had the pace been more leisurely.

A further important factor in Apollo's success, and one which played a leading part in advancing the technological spin-off from this very large scientific and engineering activity, was the complete openness with which it was conducted. It might well have been a secret and selfish endeavor. Instead, it was an open and enlightened one, devoted to the increase of knowledge for the benefit of man. Moreover, by putting the prestige and reputation of the United States on the line before all the world, President Kennedy aroused a fervor of dedication to the job in all who worked on Apollo. From top administrators and technical managers in the space agency, to scientists and engineers in universities and industry, to production workers and even floor sweepers in the plants, there was an intense feeling of personal responsibility for the success of our mission to the Moon.

But let me make clear a point not always understood. Apollo was much more than a matter of going to the Moon. The space program as a whole is far more than a matter of satisfying scientific curiosity. Science is the search for understanding of nature, and in this search we acquire mental and manual skills and knowledge that underlie the strength of society. Knowledge of scientific facts and development of technology provide the standard of living we enjoy today. The remarkable growth of prosperity here in Japan is the result of more than the hard work of the Japanese people. It is this hard work combined with the intelligent application of scientific knowledge and technological development that generates new Japanese enterprises and new jobs, creating more resources. When you invest in the creative programs of scientific research and development, you are providing for your own and future generations economic survival and well-being.

In primitive times, the main question was physical survival. Over the ages we have added a great variety of cultural and spiritual values by which we measure the quality of our lives. That these values could be added is only because science and technology have freed us of most of the drudgery of making a living.

Consider for a moment how present-day society would fare without electricity. Society would literally collapse. Factories would close. Millions would be without work. Housewives would go back to making all clothes by hand. Dentists and surgeons would revert to clumsier instruments of a hundred years ago.

Yet, it is only a little more than a century past that a scornful British Parliament asked the English physicist Faraday why he fooled around with electromagnetic fields. With a good deal of prescience, Faraday retorted that someday members of Parliament would levy a tax on the results of his research.

The history of science is the history of general disbelief by the public that the work of scientists has any "practical" value. New theories and concepts of natural phenomena, new scientific knowledge, and new ways of doing ordinary and essential things, are never associated with the opening of a new factory, say, to produce synthetic fibers and fabrics. Electric power, fast trains and faster airplanes, the dentist's high speed drill or the surgeon's cryogenic scalpel, are not recognized as once having been "impractical" ideas in the questing minds of scientists.

The work of amplifying and concentrating light into a narrow beam suddenly spins off into many fields besides optics. New techniques for cutting through thick metal plates, or precisely cutting cloth to a pattern by laser are improving industrial operations and tailoring alike. Today we can measure to within a foot the distance from the Earth to the Moon in an experiment set up by astronauts on the lunar surface. Tomorrow, with a satellite place in a stable, synchronous Earth-orbit, we can use the laser to precisely measure movements of the Earth's crust, monitor continental drift, and develop techniques to predict quakes and volcanic eruptions. Communications by laser beam will make possible another major step forward in the communications technology already revolutionized by space satellites.

Experiments with the lunar dust brought back by our astronauts show that some samples, at least, produce an astonishing effect on certain planets. They grow bigger, stronger and greener than the same plants grown in the best fertilizers we have been able to create. The U.S. Department of Agriculture's chief plant pathologist at NASA's Manned Spacecraft Center in Houston is unable to explain what it is that Moon dust contains. Yet, it does something to plants that nothing else can duplicate. If the manner in which lunar dust works on plants can be discovered, then it may be possible to imitate it, producing a great improvement in agriculture.

One of *Apollo 14*'s interesting experiments took place during the flight, not on the Moon. It was designed to show that certain organic molecules can be separated in a fluid solution. Experimenters have found that most organic molecules pick up small electric charges when placed in slightly acid or alkaline water solutions. Then, if an electric field is applied, the molecules will move through the solution.

Since different molecules move at different speeds, the faster ones outrun their slower fellows as they move toward the other end of the tube of solution. Scientists reasoned that this characteristic could be used to prepare pure samples of organic materials, such as hormones, DNA (the molecules that carry the genetic code), and hemoglobin.

But they had a problem. Gravity conditions here on Earth caused sedimentation and sample mixing by convection.

An experiment was proposed for *Apollo 14* and placed on board. It was found that the weightless conditions of space did indeed suppress both convection and sedimentation, and the desired separations took place. Further experiments will be necessary since the *Apollo 14* test was relatively crude compared with laboratory conditions. However, a dazzling new field is opening in biochemistry and pharmacology. The effects on animal husbandry may be helpful. For example, undesirable cattle genes may be systematically removed and replaced by more desirable types. Cattle breeding would be a more exact art than is possible today, and very likely results would be quicker.

These are only random examples of how Earth-based science and space science expand from ideas or chance discoveries into what we call "practical" results. Normally, if we can't understand a new concept, or attach a monetary value to it, we regard it as "impractical." Many people today do not understand the value of the space program. People generally view it as producing rather fantastic hardware in the shape of vehicles that carry men to the Moon. To them, the whole thing seems utterly impractical.

But the contributions being made by the space program that are most significant are not the fantastic spacecraft hardware, but the "software." This software is knowledge, the new concepts of nature and man, the new ways of doing things. The space program is an activity that probes into the cosmic design to bring about an understanding of its events, its laws and principles. It searches into the basic "building blocks" of life to comprehend a phenomenon that still eludes us, but in the process we are discovering things never before known about the physical basis of life. We have even discovered and demonstrated a theory that may help explain the source of uncontrolled malignant growth, or cancer. A NASA space scientist, studying the effects of radiation in space on the blockage of cell division, developed the theory. It has been experimentally verified in laboratory tests at our Langley Research Center.

Dr. Clarence Cone's theory proposes that the division of body cells is controlled precisely by the pattern of ion concentrations on the surfaces of the cells. The pattern is formed by the electrical voltage that normally exists across cellular surfaces, and varies from one part of the body to another.

Dr. Cone's theory offers, possibly for the first time, an explanation of the functional connection between two major features of cancer—uncontrolled growth of cells and spread of the disease in the body. In his research he noticed that cells having large negative membrane voltages seldom if ever divide. Cells with small negative electrical potential, however, divide at maximum rates. Dr. Cone believes there is a central mechanism controlling body cell division. If his belief is proved valid, it will provide a new basis for research progress on many biomedical problems, including birth defects, growth, aging, human conception and, especially, cancer.

Dr. Cone's work illustrates a point I wish to make, for his is only one of thousands of discoveries and innovations being made in the space program every year. Cancer research has been going on for many years. But the interesting point is that this entirely new theory on the disease was made in the course of research for the space program, not by specialists in cancer. The American Cancer Society thought it so important that Dr. Cone was invited to tell about his work before the society's Annual Science Writers' Seminar.

The space program is in fact a huge generator of new scientific facts, new theories, and new technology. These will form the basis for future medical advances, new industries, and new capabilities that will help carry forward the necessary economic development to support a growing population. Some of these innovations are in fact already being converted to industrial use in such diverse fields as shipbuilding (a light portable router to machine subassemblies) and plastic manufacture, among many others.

The space program's role as a generator of quantities of new knowledge and technology through research and development here on Earth is one side of its software production. Another is the scientific investigation of the solar system, including Earth, and the universe. We have learned a great deal about the important Earth-Sun relationships that play a dynamic role in our atmosphere, the continents and the seas.

The physics and astronomy program is a coordinated effort to study the Earth's space environment, the Sun and its role, interplanetary dust, meteors, comets, the stars, and the basic nature of the universe. It is a search not only of the origins and evolution of the universe, but of man and his place in the scheme of things.

None of this would seem to have much connection with our lives here on Earth. But it was study of the Sun that led to understanding and controlling the power of the atom. Scientists believe that study of the incredible energies of very hot stars may help reveal the secret of controlling such things as nuclear fusion reaction for production of cheap commercial power, cheap enough even to convert sea water to fresh water and at the same time eliminate the serious pollution problems of present power stations.

The scientific exploration of space and the planets by unmanned instrumented vehicles, and manned exploration of the Moon, are producing vast quantities of new knowledge. A Stanford Research Institute report said no single endeavor or phenomenon has increased the discovery of scientific facts and theories as much as the exploration of space. The report added that in the near future those discoveries will exceed all prior scientific knowledge.

But there is still a third role the space program plays in generating software, and this is our Earth applications program. The greatest resource in space is knowledge. But in acquiring knowledge, we learn how to turn it into useful services for mankind.

Among these services is the well-known weather satellite program. NASA is working to improve the technology so that eventually forecasts can be made up to two weeks in advance. This is no easy task. Weather systems are very complicated, requiring vast quantities of data to be acquired, processed and analyzed within short times.

Volcanic eruptions might be predicted by using a combination of instruments placed in the crater cone, equipped with a transmitter, and a space satellite. Volcanoes on the average erupt about every 100 years. A tilt meter placed in the crater would detect the very slight variation in the cone when pressure in the magma chamber far below builds up to another eruption. The satellite would monitor the tilt meter each time it passed over. If there was significant variation in the assigned digital call figures, this information would be transmitted to a central receiving station. The volcano could be identified and located and the necessary warning of impending eruption could be given to the communities concerned.

This has been necessarily a very brief view of some of the contributions made by such a large scientific and engineering project as the space program. Today, spacecraft are enriching mankind's knowledge of Earth and the universe. They are contributing to less-costly and better communications between all men. Every year thousands of lives are saved by timely warning of destructive storms. Just last year in the United States Gulf Coast, some 50,000 people were saved by satellite warning of the giant storm Camille. New standards of excellence set by the Apollo Program have inspired a critical reevaluation of the performance of both man and machines.

The very diversity of these contributions alone is significant. Yet, we have only begun to reap the benefits that derive from huge engineering projects. The space program has changed the course of mankind by opening great new enterprises and new worlds in space. It is up to the human race to accept the great opportunities that lie before us.

*** END ***

– Chapter 42 –

Lee Ranck, who was associated with the Methodist Church in the Office of Methodist Information, arranged this interview with von Braun and two students.[97]

What is ahead for the space program? What are the benefits of the program, and has it been worth the money spent?

These and other questions are explored with Dr. Wernher von Braun by two senior high young people for Face-to-Face. Dr. von Braun, a driving force in this country's space program since its beginning, is Deputy Associate Administrator, NASA. Jackie Ennis and Robert Ackerman, seniors at Bowie (Maryland) High School, interviewed Dr. von Braun at NASA headquarters in Washington.

Face-To-Face: 1971

Bob: WHEN THE APOLLO MISSIONS ARE COMPLETED, WHAT FURTHER PLANS ARE there for the space program?

Dr. von Braun: We will fly three more flights to the Moon: *Apollo 15, 16,* and *17.* After that we will establish a space station in orbit called *Skylab.* It will be visited by three subsequent parties of three astronauts each in 1973. These will be scientist astronauts who will do medical research on how man reacts over an extended exposure to zero gravity in orbit. The *Skylab* will also be equipped with a manned solar observatory, so part of the time will be used in observation of the Sun. The final visit to *Skylab* will be for two months, and quite a number of other things will be done, including a survey of Earth resources from orbit, using sophisticated photographic techniques, and more medical and astronomical work.

Jackie: Will this ultimately lead to transferring population to the space station to live there permanently?

Dr. von Braun: I do not believe that this will be the most important contribution of space stations. But I do believe that with the resource surveys made from the space stations we will be able to find more resources on Earth to feed and find jobs for more people on Earth. I think the Earth can feed easily twice or even four times as many people as it has today. But we have to turn away from our present method of exploiting the limited resources of the planet Earth and begin to manage them. That means we have to first find out the location of undiscovered resources. For example, there are undoubtedly vast stretches of the Earth that have oil underneath that we do not even know of, or metal ores that have never been prospected for or mined. I think in this way space stations and observations of the Earth from space will contribute a great deal to the solution of the population explosion problem.

Jackie: The orbiting space stations that are planned, will those carry nuclear weapons?

Dr. von Braun: No. There is an international treaty against the carrying or use of nuclear weapons in outer space. The United States is a signatory to that treaty; so is the Soviet Union.

Bob: Many people decry the overuse of government funds on the space program, claiming domestic issues should be the priority. How do you feel about the future of the space program?

Dr. von Braun: Most domestic issues cost money; for instance pollution control or new means of mass transportation. That money must come from somewhere and that somewhere can only be a very vibrant American economy. Now, the space program contributes a great deal to the growth of our economy and the generation of wealth. In this respect the space program is not at all in competition with the social programs, but, as I see it, is really helping create the means with which the social problems can be solved. I think it is absolutely erroneous to say that the space program has made America poorer; it has made

America richer. If we now say we will also clean up our rivers and air or build a mass transit system between Washington and Boston, we can rest assured that this will be at least partially financed out of new tax revenues generated by the space industry.

There are quite a few space activities outside of NASA that are a direct result of the NASA success story. An American corporation called COMSAT (Communications Satellite Corporation), provides telephone service via relay satellites across the Atlantic and the Pacific and enables Americans to watch European television programs. This is a money-making corporation. Associated with this corporation is an international consortium of 77 nations whose telephone companies tie into the COMSAT satellite relay system. Through this international hookup you can place telephone calls to almost any place on Earth by way of satellite—and for a lot less money than in the past when all calls had to go through submarine cables. Another example of successful use of operational satellites is the weather satellites. NASA has developed meteorological satellites that enable us to improve our weather predictions, but they are now used by the Weather Service.

Bob: In an address you indicated that NASA officials were concerned because many people equate the Apollo lunar landing missions with the total space program. You said "many see them as an end rather than a beginning." If this is a beginning, where is the end?

Dr. von Braun: The Apollo program had an objective of sorts, namely landing a man on the Moon in the 1960s and bringing him back alive. But in a broader sense, that mission set only an objective around which a new unprecedented capability could be developed—a capability as President Kennedy put it at that time, "for man to learn to sail on the new ocean," the ocean of outer space. Now that we have been several times to the Moon and back, we know how to sail on the ocean of space and would like to put this capability to use. Not only for science—where the new space capability offers us much new insight—but also for the direct practical benefits of man, through surveys of the life-supporting resources of the Earth itself.

Bob: Why is it important, in your opinion, to stay ahead of the Russians in our space program?

Dr. von Braun: I think man, ever since he came crawling out of the cave, has wanted to explore the world around him and to improve his own lot. He has always wanted to avail himself of all opportunities to do these two things. When a man stops using the opportunities he has, he is getting old, and he will soon be replaced by a younger man. The same goes for nations. An old nation is a nation that doesn't avail itself of the opportunities it has. A young nation is one that does, I think. America is not that old. It is in this sense that we ought to stay ahead of the Russians.

Bob: Do you think this sense of competition is good? Or do you think we should work jointly with the Russians?

Dr. von Braun: The relation between our space program and that of the Russian's includes both cooperation and competition, and I think that is just as it should be. It is really, in a way, the same as in our industry; between General Motors and Ford there is a lot of competition, but when it comes to developing safety standards for automobiles, they sit together and compare notes. We have a number of cooperative space programs with the Soviet Union. One is in the field of data exchange from weather satellites. We have a communications arrangement with the Soviets under which we receive every day satellite pictures of cloud formations and storms taken by Soviet meteorological satellites. They are piped into our Goddard Space Flight Center in Maryland. We call this the "cold line to Moscow." The Soviets, over the same line, get our weather data in return. Very recently a discussion was held where both sides agreed that the quality of this data transmission system ought to be improved.

There have been discussions recently in Russia and in this country on the possibility of American spacecraft visiting Soviet spacecraft in orbit and vice versa, so that a mutual rescue capability could be developed and astronauts and cosmonauts could visit each other.

This matter is still under study by both sides to determine whether it is technically feasible, how much it would cost, and when we could do it.

We have one fundamental principle in all of these joint undertakings with the Soviets: the basis of reciprocity. There simply has to be a reasonable degree of mutual give and take. We will not engage in any cooperative agreements where we do all the giving and they do all the taking. Needless to say, the Soviets expect the same.

Bob: You mentioned many advantages of the space program. Do you think they are really reaching the great majority of the people?

Dr. von Braun: I believe the majority of the people will benefit greatly from the space programs. As I said, to clean up a slum or to build a highway you need money. To create money you have to have an active and vibrant industry. The space program's contribution is often not a direct one but an indirect one. We in NASA don't rebuild slums, but we help the American economy to flourish by the new technology we develop. The total amount of money available for social programs is directly related to the wealth of the economy. There is no better way of helping clean out the slums than keeping the economy vibrant.

The implication of your question seems to be: Don't you occasionally feel bad about spending money in space rather than spending it in the slums? The answer is emphatically, no! We in NASA absolutely do not feel that we are doing something bad, or something we have to be ashamed of. We know that we are not depriving the poor and needy of anything. On the contrary we are proud of our contribution to the overall economy, which must remain strong to solve our social programs. We think nothing would be more disastrous for the country than to drop wealth-generating and know-how-creating programs such as the space programs and let the economy go to pot. The result would be that we couldn't take care of our social programs either.

Jackie: Are there any non-white astronauts?

Dr. von Braun: At the moment the answer is no, but let me put a qualifier on this right away. In the early phases of our manned program, including Mercury, Gemini, and Apollo, we had to establish extremely stringent professional standards for who should be an astronaut. For example, we said we needed experimental test pilots with several thousand hours of jet flying experience. This washed out all the ladies, and we did not find any black experimental test pilots. With the very difficult things you have to do in flying to the surface of the Moon, you can't settle for anything but the highest professional aviator qualifications to begin your astronaut training.

But we are about to enter a new era; with our new reusable shuttle vehicle we will be able to carry people from other walks of life into orbit. For example, meteorologists will do meteorological work up there, astronomers will work in orbital observatories, and medical doctors will do research in orbital laboratories. The shuttle will open up space flight to anybody who is physically fit to ride an airliner. That includes people of all races, men and women, and even young people. I think space will no longer be an exclusive arena for astronaut heroes in the future. Future astronauts will sit in the flight deck of the shuttle, but in the back you will find all sorts of people. Sizable research stations on the Moon can be supported with such a transportation system, and I think it may very well be that in one hundred years a thousand people may be living on the Moon.

Bob: Do you think inter-galaxy travel will ever be possible?

Dr. von Braun: You are talking about travels outside our solar system? I think we must distinguish here between the question of whether we have an engineering solution to this problem or whether we can intellectually define the requirements and the nature of such a trip.

The answer to the first question is that we do not know today how to build such an interstellar ship. But I could very well imagine that one day man will find ways and means of doing it.

As to the second question; to identify the problem in scientific terms we must first consider that the nearest fixed star is about four light years away from Earth. This means it takes light four years to travel to Earth from that star at the speed of light, 186,000 miles per second! The Moon, for comparison, is one and one-fourth light *seconds* away from Earth. The planet Mars is thirty light *minutes* away. So we are talking about a quantum jump of several orders of magnitude, when we talk about stellar travel.

From these vast interstellar distances alone it follows that we must be able to travel at a speed approaching the speed of light to make such a trip in a lifetime. Einstein's famous Theory of Relativity decrees that you cannot move faster than the speed of light because the amount of energy you must apply to a mass approaches infinity as that mass approaches the speed of light. You can nevertheless accelerate a spaceship up to, say 90 or even 99 percent of the speed of light, provided, of course, you have enough propulsive energy to do that. At such a near-light speed, you have a very interesting phenomena working in your favor. It is called time dilation. According to Dr. Einstein, time elapses at a slower rate as you build up your travel speed. In this way a man, in his lifetime, could theoretically travel to a star a thousand or even ten thousand light years away and return to Earth. For him time would elapse at a slower rate than for the people who remained on Earth. For example, while he was traveling to that distant star and back he may have aged thirty years. On Earth, however, time elapsed would have been a couple of thousand years. As a result, he would appear to the people on Earth as a man who departed a few thousand years ago. This is only one of the many vexing theoretical problems connected with travel to other stars.

Bob: You were involved in developing the V-2 rockets in Germany in World War II. This touches on a subject of philosophy. How does a scientist deal with questions of morality as he develops weapons to destroy?

Dr. von Braun: This is a grim problem every engineer or scientist working for the military has learned to live with. People have come to different conclusions as what their ties are. My own belief is that when your country is at war and calls on you or drafts you, whether as an infantryman or as an engineer, you have to do your duty. This problem is as controversial today in this country as it was in World War II in Germany. In Germany, many people said, "I won't do it; I will rather leave the country." Other people took the position: "This is my country and I will just have to do it." I am still of the opinion that a man has a duty towards his country in his professional field or as a soldier.

Bob: If you believe the leaders in your country are wrong, do you believe you should support them?

Dr. von Braun: Many people ask themselves that question today in this country. I think in the last analysis everybody has to live with his own conscience. All I can say is that if I were a young man today in America, disagreeing as I do with the continuation of the war in Vietnam, I would still serve in the Army. I would see this as my duty and I would not consider it morally right to defect. If other people come to other conclusions, I will still respect them. But they, too, will have to live with their conscience.

*** END ***

– Chapter 43 –

New Answers—New Questions
Texas Tech, Lubbock, Texas [98], March 20, 1972

There was a happy time—perhaps a century ago—when many persons sincerely believed they knew just about everything that was worth knowing.

Today, one of the few things we know for sure is that we still have a great deal to learn.

The more we find out—and we are finding out quite a lot—the more we need to know.

I think the story of Gertrude Stein's last words is very appealing. You remember how it goes: when asked by her friend Alice B. Toklas, "What is the answer?" she replied, "What is the question?"

In the space program, we are gathering a multitude of answers in a multitude of scientific disciplines. We are also uncovering a lot of new questions. We find ourselves in a situation similar to that of the famed astronomer-philosopher Sir Arthur Eddington in the late 1920s. Discussing the theory of atomic structure, Eddington wrote, "Something unknown is doing—we don't know what—that is what our theory amounts to."

Sir Arthur was not the least bit discouraged about the situation—quite the contrary. We can profit from his attitude and his example. Even in his very scholarly work, *The Nature of the Physical World*, he never lost his lighthearted touch.

Not many writers, for example, would dare try to explain atomic theory by illustrating with Lewis Carroll's well-known lines, "The slithy toves / Did gyre and gimble in the wabe."

Eddington was making the point that when we can quantify things—even unknown things—in known numbers, we begin to achieve real progress. "By admitting a few numbers," he stated, "even 'Jabberwocky' may become scientific."

As an aid in visualizing the role of the electron in changing states of the atom, he wrote, "To contemplate electrons circulating in the atom carries us no further; but by contemplating eight circulating electrons in one atom and seven circulating electrons in another we begin to realize the difference between oxygen and nitrogen. Eight slithy toves gyre and gimble in the oxygen wabe; seven in nitrogen. We can now venture on prediction; if one of its toves escapes, oxygen will be masquerading in a garb properly belonging to nitrogen. In the stars and nebulae we do find such wolves in sheep's clothing which might otherwise have startled us."

And he concludes, "Out of the numbers proceeds that harmony of natural law which it is the aim of science to disclose."

My subject here is supposed to be space, not nuclear physics, so I must reluctantly leave Sir Arthur Eddington. But I do think we can draw some parallels.

For one thing, I don't think anyone today seriously subscribes to the once honored theory that what we don't know won't hurt us.

We have learned from experience that every time there is a new scientific breakthrough, in the space program or anywhere else, we are confronted with what I might call the "new answer—new question syndrome."

With each breakthrough, there is a gain in knowledge; there is promise; there is a new door opened before us.

But we cannot predict what lies behind that new door. There may be not just new knowledge, but also new dangers, new problems, new responsibilities. There can be hidden or unpredictable social or other costs.

But we cannot and must not let our fears of such possibilities stampede us into closing and locking all the doors, as some today would have us do.

It would be foolish to deny that technology can bring with it unexpected problems, but it is equally foolish to think that a return to the Stone Age will solve everything. Let us recognize that technology has to be, above all, responsible. We must use it for salvation, not ruination.

I am convinced, for example, that today's sudden widespread interest in ecology and the preservation of our planet owes a great deal to our new view of the world as seen from space—particularly, the magnificent picture of the entire Earth that was immortalized several years ago on a postage stamp. The legend it carried, you will recall, was "... In the Beginning." The words now seem doubly appropriate, for reasons that should be obvious.

But the space program has taught us many other new things about our Earth and its environment. We've had worldwide looks at the weather; we've viewed Earth's resources not just in visible light but in the invisible portions of the spectrum, learning a great deal in the process—and, as usual, raising some brand new questions.

We have been reminded of how many serious gaps remain in our body of essential knowledge about our home planet. We have no idea, for example, how stable its present climate is. How much more will it take in the way of atmospheric and water pollution to upset the balance and perhaps alter the climate drastically? We really don't know. But space research will help us find out.

We know that the space program offers one of the better ways of quantifying many things we previously couldn't even begin to measure. And that, I would submit, is in accord with Eddington's basic philosophy.

Similarly, we are gaining new answers—and new questions—about the Moon. For all we really knew a few years ago, our nearest celestial neighbor might actually have been made of green cheese. Reputable scientists were theorizing that the lunar surface was covered with what they described as "fairy castles" of dust, to a depth of perhaps 50 or 100 feet. A few even direly predicted that when our Apollo spacecraft made its first landing it would disappear into the depths and never be seen again.

That particular fear was allayed when our unmanned Surveyor spacecraft soft landed on the Moon. Other unmanned craft, including Rangers and Lunar Orbiters over a period of several years before Apollo, answered a number of other questions—and raised a few new ones.

The whole world watched as our *Apollo 11* astronauts landed on the Moon and brought back incredible amounts of new knowledge about the surface of the Moon in the immediate area of the Sea of Tranquility. Later Apollo flights have done the same thing for other areas of very different history and topography. But—excuse the same old story—they have also raised a whole galaxy of new questions.

Apollo 16, which heads for the Descartes highlands of the Moon next month, may answer some of these questions. So will the final Apollo mission, *Apollo 17*, when it is launched next December. The combined data from all the lunar exploration missions will undoubtedly keep our scientists busy for the next ten years. And a good thing, too, since if there are any more Moon walks during the rest of this decade they will be taken by cosmonauts, not astronauts.

New answers—new questions. We have learned a great deal about the mysterious planet Mars from our most recent Mariner mission, once the dust storms finally died down. Now we are looking much deeper into space, out toward the edge of the solar system.

People have not paid much notice to the recent launch of *Pioneer 10* on its deep space and fly-by mission to Jupiter—probably, I imagine, because it is not due to arrive until December 4, 1973. This built-in and inevitable time lag takes away a great deal of the drama of the mission. In my opinion, however, this venture is one of the most exciting—and in many ways the most ambitious—of NASA's unmanned missions to date.

The Pioneer spacecraft must run a veritable gauntlet of hazards to reach the giant planet Jupiter. It will pass through the whirling space debris known as the asteroid belt, and then must face intense belts of trapped radiation around the planet itself.

Jupiter is so far from Earth that it will take radio commands and the returning data they trigger a round-trip time of 90 minutes to traverse the distance. NASA's Deep Space network will be put to its hardest test to pick up the signals, which over that distance will have a strength of only 10 -17 watts.

Still another problem is that the spacecraft will be so far from the Sun that we cannot use solar energy for power; *Pioneer 10* thus is our first deep space probe to employ nuclear power. It carries four SNAP-19 radioisotope thermoelectric generators, fueled with plutonium-238. These units supply a total of 160 watts of electricity to operate the experiments and the telemetry system.

For the engineers in the audience I am most pleased to report that the flight so far is going extremely well. The first midcourse maneuver was performed on March 7, by means of a relatively small velocity change of 14.5 meters per second. *Pioneer 10* is now almost exactly on its prescribed course, and any further correction, if necessary at all, should be extremely minor.

You might be interested to know that the maneuver was performed in a manner different from the typical method used in, say, our Mariner planetary missions. Instead of a preprogrammed maneuver which goes to completion once started, *Pioneer* was first precessed to the desired direction. Then the thrusters were pulsed until the desired Doppler change was observed to have occurred. (At the time, two-way light time was about 50 seconds.)

As of last week, a total of six of the on-board science instruments had been put into operation and are working well. A device to detect meteoroid "hits" has reported several small ones so far.

The instrumentation is programmed to take a reading on the numbers and sizes of the fragments making up the asteroid belt I mentioned previously. It will take *Pioneer* about 180 days travel time from date of launch to reach the nearer edge of the belt, which lies between 170 and 345 million miles from the Sun.

When *Pioneer* reaches the asteroid belt, it will deploy four telescopes to measure light scattering from the particles that make up the belt. Preliminary estimates indicate there may be about 50,000 bodies in sizes ranging from one to 500 miles in diameter. There are undoubtedly hundreds of thousands of smaller fragments, and literally billions of dust particles.

Fortunately for the spacecraft, this material is dispersed quite thinly, and with any kind of luck we hope that it can avoid any major collisions.

At the same time *Pioneer* is taking its readings in the asteroid belt, the spacecraft's magnetometer and other experiments should be gathering new information on the characteristics of interplanetary space.

As the craft approaches Jupiter, eventually the pull of the planet's gravity will accelerate *Pioneer* to four times its approach velocity.

The trajectory is planned to carry the spacecraft to within one diameter of Jupiter—that is a distance of 87,000 miles. At that point the mission becomes even more interesting and exciting, because the craft should be well inside Jupiter's radiation belts. Our best estimate at present is that these belts may be something like one million times as intense as the Van Allen Belts that surround our own planet.

Frankly, we will not be too surprised if the particle bombardment knocks out *Pioneer's* instrumentation at that stage of the mission, bringing our data transmission to an end. But if we are lucky and it keeps working, we should get a good look at the planet's surface, survey its magnetic field, and perhaps get a close-up look at one of the planet's 12 Moons.

The actual fly-by will last for roughly a week, the first two days of which will offer *Pioneer* a planetary view in full sunlight. As it draws nearer, more and more of the surface will be in darkness, and at the nearest point about 60 percent of the planet will be dark. Eventually the spacecraft will pass out of Earth view for a time as it goes behind the planet. If it is still transmitting when it reemerges, we hope to get an occultation reading of the planet's atmosphere.

Beyond Jupiter, our experimenters are optimistically hoping that *Pioneer's* instruments may measure the limits of the solar wind—or to put it a bit more elegantly, the boundary of the heliosphere. Some scientists predict that Jupiter may have its own magnetospheric "tail," just as Earth does.

Plans call for observations to continue for at least three months after the Jupiter fly-by, to study galactic cosmic rays, interstellar hydrogen/helium rations, and plasmas originating from bodies other than our own Sun.

I think you can see from all this why we at NASA are so excited about the mission. But to return to my theme, the quest for new knowledge, what do we hope to learn about Jupiter itself? What are the mysteries? How do we hope to solve at least a few of them?

We know, for example, that Jupiter has one thousand times the volume of Earth. But the planet apparently is not very dense; it only has 318 times the mass of Earth, leading us to speculate its composition may be similar to that of the Sun—largely hydrogen and helium.

We know that Jupiter has a year equal to something less than 12 Earth years, and a day of about 10 Earth hours. Because the planet rotates so fast, it is flattened at its equator by nearly 12,000 miles. Its gravity is 2.36 times ours.

One of the most interesting—and mysterious—features of the planet is the famed Red Spot. I commend to your attention an article by Astronomer Gerald P. Kuiper in the February *Sky and Telescope*. He has applied studies of our Earthly cloud formations in an effort to solve some of Jupiter's mysteries. From this work he has arrived at the belief that Jupiter's Red Spot is the cloud layer at the top of a steady-state storm—what meteorologists call the "anvil" over a tropical storm on Earth.

A telescopic observation of Jupiter made in 1971 indicated that the Red Spot is actually about three miles higher than the surrounding clouds. Because of this, Kuiper believes that we can eliminate the "floating solid" theory of the Red Spot, since the latter would require a depression in the clouds.

Another highly interesting thing about the planet that we have known, or at least think we have known, for some time is that about three-fourths of Jupiter's heat apparently comes from inside the planet rather than from the Sun. Kuiper's theory fits in with this, as there are apparently no day-and-night temperature variations as there are on Earth. Accordingly, once established, a convection pattern may continue indefinitely, and would explain why the Red Spot is permanent. But Jupiter rotates so fast, as I previously mentioned, that it has the effect of smearing the spot into an oval.

The Kuiper theory is most interesting. Perhaps in a couple of years *Pioneer* will produce the appropriate evidence to prove or disapprove its correctness.

From my brief and necessarily oversimplified discussion, you can see that we do know a few important things about Jupiter—but, oh, the things we do not know. We haven't the faintest idea what lies below the cloud tops. We know nothing whatever about the structure of the atmosphere beneath the clouds. In fact, we can't even be sure whether the planet has a solid surface.

We believe *Pioneer* will give us some answers, perhaps quite a few. We can be sure that if it does, it will at the same time raise plenty of new questions before it leaves the solar system to journey endlessly and eternally in the galaxy.

Meanwhile, back at the ranch—our own planet Earth—we will be devoting much attention to studies of its physical makeup and resources during the next few years. The first Resources Technology Satellite, or ERTS, as it is called—I have often said I wish someone could devise a more euphonious name—is due to fly in about two months. It is designed to take inventory and monitor the condition of forests and crops; detect disease and insect infestations; locate natural resources of fresh water and minerals; spot pollution of waterways and the air; supply data on oceanography and geography; and even aid in urban planning.

A second satellite in this series is scheduled for launch next year. Also in 1973 will come the launch of *Skylab*, our first experimental orbiting laboratory, which will also perform a number of Earth resources experiments, among others.

After *Skylab*, there will be a hiatus in the manned space program for about five years. And that is one of the reasons the Space Shuttle is so important to our program. For this versatile vehicle, make no mistake about it, is the key to our nation's future in space, for both manned and unmanned missions.

The Space Shuttle will be developed over the next six years. Test flights of the orbiter, flying within the atmosphere like an airplane, will begin in 1976, and manned orbital test flights in 1978. We expect to have the complete Shuttle system in operation before the end of this decade.

But what are the advantages of the Shuttle? What will it really do for us?

To begin with, it will do more efficiently the same types of useful civilian and military space missions we are doing today by providing access to space and return to Earth on a routine, quick reaction, and economical basis.

Because the Shuttle is reusable, we will no longer have to use each launcher and each satellite just once and then throw them away. What we've been doing is sort of like buying an expensive automobile, driving it one time, and then throwing it in the ocean. That way just doesn't make sense any more. We've completed the early pioneer work of the last dozen years and have learned ways of doing things better.

On this score, I might mention one additional point that is even less well-known to persons not actually engaged in the space program. Right now we are using some 17 different combinations of boosters and upper stages—all the way from the relatively small Scout vehicle to the enormous Saturn V. To launch these various sizes and configurations of vehicles, we have built a total of more than 80 launch stands in this country, and 50 of these are still active or on standby. With the Shuttle we can consolidate operations to a great extent; in fact, we can perform multiple missions in a single flight.

The Shuttle will allow us to go into space routinely, without the enormous effort and lengthy countdown procedures we need now. It will enable us to place satellites in orbit; return satellites from orbit for repair, refurbishment, and relaunching. As things now stand a minor malfunction can put a satellite costing perhaps $80 million or more out of commission and there's nothing we can do about it except launch a new one. With the Shuttle we can send up a man with a screwdriver to fix it, or if necessary bring it back down for a thorough overhaul.

With this kind of capability, we will really be killing several birds with one stone, flying all kinds of missions—manned and unmanned; civilian and military; launching communications and weather satellites, scientific satellites, monitoring Earth resources, and even performing manufacturing operations in space.

All of which, after a few detours, brings me back to my basic theme of new answers and new questions once more. The Shuttle will aid us greatly in our study of the Earth, the Sun, the planets, and the universe. One of the most interesting possibilities is that scientists will be able to accompany their own experiments into space without having first to undergo years of intensive training to become astronauts. Only the two men actually flying the Shuttle will need such training. Others will be riding as passengers.

We also expect it to encourage increased international cooperation. We have invited foreign countries to participate, either through flying foreign nationals in space, or by incorporating foreign experiments. There is even a possibility of international cooperation in various aspects of the development of the Shuttle itself.

It may well also be one of our best hopes for achieving closer cooperation with the Soviet Union. We can be far more effective in gaining the cooperation we want if we do not hand them a virtual monopoly in the realm of orbital flight—which is what we would be doing if we were not building the Shuttle.

There has been excellent progress toward increased cooperation. American and Soviet officials have had a number of meetings to talk about some of the future possibilities in this area. Working closely together, there is no doubt that both countries could do much more in space and do it at less cost—not just for the benefit of our respective countries, but for the whole world.

I do not mean to imply by all this that the Space Shuttle—or even our total space program—is going to be a panacea for the world's ills. Remarkable as they are, space systems have inherent limitations. But I sincerely believe that our space program is a tremendous driving force, a unifying force, at a time when many other things are pulling individuals and nations apart rather than together.

Mankind has been given powerful new technological tools in the search for new answers and new questions. The decisions we make, the actions we take, and the policies we implement will be clearly unfolded in the years to come. Much depends on how well we do our job.

*** END ***

— Chapter 44 —

The Goddard Symposium is sponsored annually by the American Astronautical Society (AAS), a network of space professionals dedicated to advancing all space activities. Von Braun presented the keynote speech at the Goddard Symposium at the Washington Hilton, March 13, 1972.

Benefits From Space Technology [99]
Goddard Symposium, March 13, 1972

WE ARE HERE TO EXPLORE THE RELEVANCE OF AEROSPACE TECHNOLOGY TO industry and the community. It is not an easy assignment.

One of the most difficult tasks is to demonstrate for public understanding the relationship between a vigorous aerospace industry and the national economy, between American leadership in aerospace science and technology and America's high standard of living.

The major contributions to our standard of living by the aerospace industry are not felt directly by the bulk of our population, and therefore are not identified with the industry. Even more remote are the effects of large federal programs in developing aerospace science and technology which are basic to the industry's activities.

Superficially, it is deceptively easy to furnish numerous examples of innovations from aerospace programs which have entered [the] non-aerospace fields of medicine, commerce, business, industry and education. The list is impressive. But, individually, and scattered throughout the economy, they make little or no impression on the public.

It is even harder to measure precisely the intangible gains from space activities in terms of national prestige, spirit, and life concepts. Yet the achievements in space have contributed significantly to all of these things. It is impossible to put numbers on these abstractions, but they may be the most important legacies of our activities in space.

One of the problems of informing Americans of the very real contributions made by the space program is the lack of visible, direct links between what we spend on space activities, and any important gains for society. There are literally no tangible, direct benefits that are *visible* to Americans which seem important.

We can talk all we want to, for example, about the important scientific and technological capabilities we are creating for the nation—capabilities which are basic to much of what the public wants done—and it means exactly nothing. People do not relate science and technology to the everyday business of living, fighting the daily traffic, and buying the groceries. If they do, they are apt to cuss it.

We can point in vain to communications and weather satellites which are revolutionizing worldwide telephone, television and weather forecasting techniques. People simply yawn. They rarely if ever phone Europe or Japan, they can still catch the ball game on TV, and they still get caught in snow storms.

So, who needs the space program?

In this symposium, we are addressing ourselves to a narrower discussion of space technology transfer to community and industrial activities. Yet, I would hope that some thought for public understanding of the subject matter will be included in the presentations, for that is the ultimate aim of these deliberations.

Almost two years ago, Dr. William D. McElroy of the National Science Foundation told a House committee that "civilized man cannot long survive on this planet without increased creation of new knowledge and its enlightened use."

He added that the best means to insure continued support of basic science and its smooth transition to applications were the mission-oriented agencies.

Dr. McElroy was speaking for the need of a sweeping reevaluation of science and technology and their roles in society. But he could have been describing the broad role of large NASA programs, such as Apollo and the manned Space Shuttle. NASA is a mission-oriented agency. Its programs, like Apollo and the shuttle, focus the energies, the brains and imaginations of large numbers of talented people working in all the disciplines in a concerted, coordinated effort which advances science and technology over a broad front. NASA grants have supported research for a broad body of basic studies, and our scientific spacecraft have telemetered back to Earth vast quantities of basic data on phenomena of the solar system and universe. Returns from the Apollo lunar explorations are so great that it will be years before we will know its full value to mankind, but already the increase to our knowledge and understanding is priceless.

These contributions alone go a long way to justify the space program. But the benefits to be derived are largely in the future, whereas the public is more sensitive and responsive to the here and now. Many examples exist of space technology adapted to non-space use, but the significance of almost all of this has had little visible effect in the public consciousness.

Of course, we all expected it would be a little while before our space activities brought benefits easily identified as important in the public eye. What we did not foresee among future "benefits" was that Apollo would arouse a new, more utopian concept of what government can do to improve the quality of life here and now on Earth. Demands and expectations for immediate improvements in society are on a scale which considerably exceed the cost of a manned expedition to Mars.

Ironically, in the general clamor to reorder national priorities, NASA—far from being credited for this service—found itself relegated to a level of decreased budgets.

Our difficulty is that, as a nation of short-term pragmatists, Americans are not geared mentally to long-range planning and deferred benefits from advanced science and technology programs.

The coming decade may see some shift in the public attitude and a growing awareness of the contributions of space, however, as applications of technology from NASA programs gain prominence. Space technology has a substantial value to the non-aerospace field. We all know it, but the problem is to get the public to know it. The simple fact is America has built its strength and its standard of living on its leadership in technology. In the past two decades or so, our high productivity in high technology products has kept us ahead of the competition from abroad. The only way to stay ahead is to continue to invest in our high-technology-producing aerospace programs.

A glance at our big foreign trade surpluses tells why. The excess of annual U.S. exports over imports is in high technology products, such as aerospace, running at $3.6 [billion], computers at $1.1 billion, and machinery, $1.0 billion.

Just the reverse is shown by our low technology products. Motor vehicle imports over exports are $3.3 billion, followed by clothing and textiles, $1.9 billion, and iron and steel, $1.9 billion.

I think these illustrations indicate quite clearly which side our bread is buttered on; we are strongest in high technology and weakest in low technology competition. American leadership in aerospace research and development has given us a world market in products which require constant advances in high technology and high productivity, activities in which we have excelled for a generation or more.

In the low technology fields, however, where products can be produced by cheaper labor, the competitive advantage lies with foreign industry where wages are lower. Although this country is the leading motor car producer in the world, for example, the bulk of our car market is within the United States, and export sales abroad meet stiff competition. Foreign car manufacturers, however, are increasingly building up their sales here because they are competing against labor costs, not high technology.

This is the broad picture of the benefits coming out of our aerospace programs and industry. I think it is extremely important that we Americans gain an understanding of how basic the development of high technology—especially high aerospace technology—is to our standard of living.

Every industrially-capable foreign country is ambitious to penetrate the lucrative markets our high technology has developed, of which the domestic American market is the prime target. Already there is a rising tide of competition from overseas in electronics, aircraft, instruments and controls, telecommunications, and others. It isn't hard to see where this trend will lead if we permit it. The only alternative visible—and I think you will agree that it is a most undesirable one—is for Americans to reduce wage levels so that we are competitive with overseas labor in the low technology areas.

Now let's look at some other benefits of space and space technology.

Space exploration has made a real impact on astronomy. Historically, astronomy has played a key role in the development of civilization for thousands of years, in agriculture, geography and navigation. In later times, astronomy has figured in a welter of discoveries in the physical sciences and their experimental tools, including optics, spectroscopy, specialized photography and infrared techniques. Traditionally also, astronomy has served to advance mathematics and philosophy.

Space technology has now added a new dimension to astronomy by providing better observation of celestial objects, through the elimination of atmospheric refraction and systematic errors. Benefits range from a large number of new discoveries in planetary physics, geophysics, and the chemical and physical properties of the solar system and the galaxies beyond.

All these discoveries are changing our concepts of the universe and man's place in it profoundly. Concepts of man and the universe, and man in the universe, motivate our thinking and actions on Earth. Are contributions to such concepts unimportant to the quality of life Americans strive for today? On the contrary, I think they are basic to the definition of what we mean about quality in life. Without a growing precision of our definition of the universe and its elements, we cannot hope to improve more than the physical aspects of day-to-day living; predictably, life would then soon degenerate into crass materialism.

So, by providing spacecraft to carry the astronomer's instruments and telescopes, the space program is contributing to the fundamental welfare and betterment of mankind. True, these are long-term contributions. They will not directly put food in our mouths, roofs over our heads, or clothes on our backs. But surely no one will say they are unimportant, that we can put off acquiring the abstract benefits of space astronomy indefinitely because of more "urgent" problems. Is materialism our most urgent requirement?

The correct answer, I believe, must be to supply both our material and conceptual requirements simultaneously—sufficient to the physical welfare of people and the growth of cosmology and the sciences.

In the area of technological innovation or spin-off, some of which will be discussed here today and tomorrow, there have been major civilian applications that are now in commercial development. NASA has described many thousands of other potentially commercial space program developments which are available for use. However, because the time cycles of orthodox American industry—from invention in the laboratory to the appearance of a product on the market—range anywhere from 10 to 25 years, only a relative few examples of transfers have been documented.

In the medical field, there have been individual instrument and equipment transfers that are particularly impressive. These range from improvement in x-ray diagnostics and physiological sensors to special equipment for handicapped patients, and a widening use of remote monitoring of hospital patients.

Some contributions from the space program occur when research spurs consolidation within a field by developing an application for known technology not previously used in that manner. One such contribution, in fluid dynamics, was the consolidation of technology in very low pressure devices for control systems. A machine tool manufacturer noted its significance and studies it for use as a control principle in operating automatic turret lathes. A prototype proved successful, and the company has built these machines in three sizes. They have been sold in a price range of $35,000 to $75,000 each.

The petroleum industry has been aided by technology directly resulting from space research. High-quality color photographs of the Earth taken by our astronauts have helped locate potential oil-producing sources. Navigation via satellite has allowed marine explorers to fix their positions regardless of the weather. Airborne multi-spectral scanners, developed for the Earth Resources Program, have provided color imagery of the terrain, permitting identification of different rock types, including oil-producing outcroppings. High-speed gravity measuring techniques, used in studying lunar gravity, permit faster and less expensive survey of offshore areas. The magnetometer experiments deployed by Apollo astronauts are adding to our knowledge of continental drift, which petroleum geologists find of direct interest in their work.

In addition to its contributions to liquefied natural gas carriers in shipbuilding, NASA-generated technology has opened new possibilities for the surgeon in cryogenic surgery. In surgery, cryogenics has been used to treat Parkinson's disease, remove tumors and cataracts, and for bloodless tonsillectomies. Cryogenic superconductivity signals a new generation of motors, computer memory cores, power transformers, magnets and transmission lines.

It now seems that the food industry, facing increasing problems in the use of food additives and nutritional quality, may be able to draw on NASA's extensive knowledge about food processing, preservation, and nutritional value. We have extremely rigid requirements for the food we supply the astronauts, since any type of food system failure would have grave consequences. Food must be free from bacterial contamination; it must be of high, known nutritional value; it must be stable without refrigeration under wide temperature variation for long periods of time; and it must be capable of fast, reliable and foolproof preparation.

The precautions we have taken with food for the astronauts have led to new and improved methods of processing, preserving and sterilizing it. These may have significant value to both the food industry and the consumer. For example, a flour company, responsible for a number of food items for NASA, has developed precooked, pre-buttered rolls that are preservable up to 600 days. In another case, a scientist under NASA contract produced an instant rice that is truly "instant." Normal hot tap water (about 155° F) can be used, and it takes under three minutes to serve from shelf to table.

Using NASA-developed methods, certain foods can be prepared and stored for emergency situations. If a disaster occurred, the food could be immediately shipped to the stricken area with no loss in preparation time and no need for refrigeration.

Space research and aeronautics, as we all know, have a lot in common. Aeronautics has contributed a great deal to the advance of space systems, and similarly the space effort has reciprocated. One of the most notable examples that comes to mind is the space program's contribution to air navigation.

Before the airlines began populating the skies with 10-mile-a-minute jetliners, precise navigation—while important—was not so critical as it is becoming today. Now, however, thanks to Apollo, airline pilots have available a new navigation aid that gives them instantaneous and continuous position reports. It operates on the same basic principle that enables our Apollo astronauts to pinpoint their positions far out in space. The heart of this system is a computer fed by a series of accelerometers which sense every movement of the airplane, up or down, sideways or forward. The computer translates this data into the instant and continuous position reports required by pilots flying the crowded airways.

The system is self-contained, and therefore independent of radio, radar, weather, and interference. Because it is more accurate than previous navigation devices, it contributes to a shorter flight time and savings in fuel. Only the most modern airliners now have these nav aids, but the new generation of aircraft being built will employ them widely.

Related to this benefit is another planned navigation aid of an even higher order. NASA, together with the airlines and other government agencies, has used the applications technology satellites to demonstrate how a satellite may serve as a space reference point.

We have shown that the satellite can provide position fixes of exceptional accuracy. The planned operational system will be a complementary means of navigation to individual aircraft. In contact with both airplane and land-based stations, the satellite can give the traffic controller an independent means to follow the airliner's progress. Satellite-relayed data can confirm or correct the pilot's position reports, which are based on his onboard equipment.

Today there is a profound concern about clean water and clean air. Space technology can play an important role in helping achieve both. Environmentalists may be interested in an advanced type of sensor that measures carbon monoxide concentrations; this device was developed in a program initiated by our Langley Center. The Langley experiment is designed to make global measurements of carbon monoxide over a period of a year in order to map the portions of the atmosphere with high, low and average concentrations of the gas. In this way, scientists hope to identify the so-called removal sinks in which the gas is changed to another compound.

This may help to solve the riddle and explain why the total concentration of the gas in the air is not increasing, although motor car exhausts, industrial activities and other sources generate some 200 million tons of carbon monoxide each year. Scientists have estimated that the atmosphere contains about 500 million tons of carbon monoxide, but measurements over a period of years have not shown appreciable increases. This implies there must be some natural mechanism that removes most of the gas as it is generated. It would aid environmental control efforts to know what the mechanism is and its capacity for converting carbon monoxide to another compound.

In the effort to obtain cleaner water, we have a lightweight precision sensor designed to detect color gradations in water so oceanologists can spot pollution. Called a multi-channel ocean color sensor, it is more sensitive to color than the human eye, and will be flown initially on aircraft. Eventually, it may be placed aboard spacecraft for monitoring the oceans. Data it provides can also be used to spot areas where fish are likely to be, and to study marine biology.

Given repetitive exact color information from this device, oceanologists can chart trends in pollution and marine life for use in preserving and developing ocean resources.

The major contribution for environmental uses is, of course, the Earth Resources Technology Satellite, now scheduled to be launched in the first week of June. We believe this may have the greatest potential of all for realizing hard economic returns from space exploration.

Information from the ERTS battery of sensors will be relayed to an Earth-based, computerized data handling and analysis network. In this way, regional data banks all over the world will receive daily volumes of data that can be put to work for man's benefit in three basic directions:

> The information will help provide more of everything through far better management of the world's resources.
> It will uncover new resources.
> It will identify trouble areas so that remedial action can be taken at the earliest possible time.

To give you an idea of the volume of data traffic to be received, the ERTS satellite will relay several hundred million bits of information daily to Goddard. Goddard in turn will produce some 300,000 color, black and white, and digital tape photos of the Earth's surface each week.

Our *Skylab*, to be launched next spring, will tie into ERTS experiments by supplying complementary information from astronaut-monitored experiments aboard the space station. At the end of each manned visit to *Skylab*, the astronauts will bring back with them data on photographic film and digital data on magnetic tape. These *Skylab* experiments will form an important part of the development of instruments and sensors for the ERTS program because the astronauts will monitor their performance.

The space shuttle, however, offers the greatest potential for space benefits over the long term. It will be the first true space transport, due to its versatility as a carrier of men and equipment, its flexibility in operations, and its ability to make repeated missions on a routine basis. While the civilian-oriented missions are the most intriguing in NASA's eyes, the Defense Department also sees its usefulness in military operations—for the launch and recovery of surveillance satellites, for example.

The shuttle's cargo bay can be used not only to carry a variety of spacecraft to be placed in desired orbit, but also will accommodate a complete laboratory module in which scientists and engineers can conduct experiments under space environmental conditions. In addition to placing spacecraft in orbit, the shuttle can retrieve them for repair, or to install new experiments, and reuse. This means that scientific and other space satellites can be built less expensively than they can at present, because designs need not include such high standards for reliability. More off-the-shelf items can be used, resulting in large dollar and time savings.

The space shuttle now planned will be only the forerunner of much more advanced multiple-mission type spacecraft. The knowledge we shall gain from its design, construction and operational performance will teach us a great deal more about such vehicles and their uses. To fully utilize space for the benefit of man, the multiple-mission spaceship is a must.

The work that NASA is doing in advancing science and technology—pushing forward the frontiers of knowledge—cannot be overstated. The vigor with which we carry on the development of science and technology, the spirit with which we explore the unknown, are uncharacteristic of vital, growing nations that will be ready with the answers to problems of the future.

There's a little-known anecdote from history which illustrates the role played by science and technology in the life of a great nation. It appears in Volume 4 of Joseph Needham's *Science and Civilization in China*. It relates how political decisions dimmed and finally extinguished the final blaze of splendor of one of the world's great civilizations.

Needham says that a series of seven expeditions began to explore the seas to the south and west of China in 1405. Under command of Cheng Ho, the fleet carrying 37,000 men reached the town of Malindi, in what is now Kenya, three-quarters of a century before Columbus made his first voyage to America. By 1433, Cheng's fleet had reached Mecca, and by the time he left Africa for the last time the Portuguese had hardly begun to explore the continent's west coast.

Compared with contemporary European ship technology, the Chinese armada was a revelation. Sixty-two of the ships were 9-masted galleons, 450 feet long from bow to stern and more than 180 feet in beam. The vessels not only had multiple masts, but also fore-and-aft rigged sails, and true axially mounted rudders. Strong bulkheads divided the hulls into naturally watertight compartments. The Chinese also installed pedal-operated bilge pumps, developed from the principle of the noria, used in ancient China to raise water into irrigation channels. These last two items did not appear in European ships until the 18th century.

Beside these great ships, Columbus's vessels appear pitiful.

Cheng Ho's expeditions were, however, the last important explorations of the seas made by China. The political decisions that killed them, Needham says, were part of a decisive *turning inward* of China's civilization. Despite China's early and extensive lead over Europe in science and technology, it was not in China that the scientific revolution took place. Yet, a Chinese visitor to England in 1400 A.D. would have considered that country technologically backward.

Let us hope we can learn from history. There have been disquieting signs of an "inward turning" among Americans which we can only hope is but temporary. Carried to the extremes that occurred in China of the middle ages, it could be nothing less than a catastrophe to modern America.

Scientifically and technologically, however, we have built the foundations for America's greatest age. Space offers the opportunity to fulfill the requirements of the spirit while we increase our capabilities to meet the material needs of ourselves and fellow men.

*** END ***

– CHAPTER 45 –

Von Braun recognized the "tremendous potential" of communication through satellites in Earth orbit for many years and he used every opportunity to promote the idea of using orbiting spacecraft for that purpose. Others joined this campaign, most notably Sir Arthur C. Clarke, who is credited with the original idea of putting a communications satellite into an equatorial geosynchronous orbit where it would remain above the same point of the equator indefinitely. Other countries were almost as quick in realizing the tremendous potential of communication satellites. The Soviet Union was the first, bridging the vast domestic distance between European Russia and Vladivostok with its Molniya satellites and the first country to acquire its own domestic synchronous satellite was Canada.[100]

REMARKS BY DR. WERNHER VON BRAUN
APPALACHIAN EDUCATIONAL SATELLITE PROJECT [101]
AUGUST 23, 1974

IT IS A GREAT PLEASURE FOR ME TO BE HERE WITH YOU THIS AFTERNOON, AND TO help you mark a truly historic moment in the history of communications satellites. You, as professional teachers, have become the first educators to complete in-service training via satellite. So, you are really space age teachers. I congratulate you and commend you for investing your time and talents in this training in order to improve your effectiveness with your own students.

There will come a time, I suppose, when your students will take this satellite teaching for granted. That is the fascination of space communications. The technology of satellite transmission is advancing so rapidly that what one generation marvels at, the next generation discards as obsolete. It is fascinating to do some crystal ball gazing and speculate on what those developments will be that will cause our children—the next generation—to marvel at. For the next few minutes I would like you to join me in a look into the future, to see how teachers such as yourself, and all of us, will use space communication ten or twenty-five years from now.

We can begin, of course, by looking at the ATS-6—the new satellite that carried the educational instruction which you have just completed. ATS stands for Applications Technology Satellite. As its name implies, it is a communications satellite designed to use all of the technology we have learned in practical applications to improve the way we here on Earth communicate with each other.

You have seen, first hand, how ATS-6 is powerful enough to beam a television signal directly into low-cost, simple antennas hooked up to your television sets. ATS-6 is the first satellite powerful enough to do the job without the need for the huge, complicated ground receiving stations that have been needed with all previous satellites. For the first time, people who live in remote places, some with no television at all, will now have space age television, via satellite. Besides serving the Appalachia area, ATS-6 is also helping people in the Rocky Mountain States, the Northwest and Alaska. In addition to giving in-service training to teachers such as yourselves, and direct instruction to their students, ATS-6 will allow people in isolated communities to get better medical care. A doctor hundreds of miles away will be able to see and talk to his patient, listen to his heartbeat, evaluate an EKG and prescribe a course of treatment to the local health aide, all via satellite.

There are many other applications of this new communications satellite. Many of them are too technical, and I won't put you to sleep with a lot of engineering talk. But take a quick look, with me, at some of the other exciting things this satellite is doing. ATS-6, for example, is communicating with other satellites. Because of its great height above the Earth and its stationary orbit, which keeps it fixed over the same spot, ATS-6 can see a much greater part of the Earth than low-flying weather satellites. The weather satellites, because they are in a low orbit, are out of touch with ground stations much of the time. To solve that problem, information gathered by a weather satellite will be relayed up to the higher ATS-6 communications satellite and then to the forecasters on the ground. The result is that satellite weather information gets back to the weatherman much sooner. In setting off the alarm for violent weather, sometimes saving minutes means saving lives.

In another area, the ATS-6 communications satellite will experiment in better ways to maintain the airspace between airliners flying over the oceans. Since communications with airliners far out over the Atlantic are now spotty, air traffic controllers must allow a great deal of distance between aircraft. With more efficient communication satellites, such as ATS-6, controllers will know the real-time exact location of each airplane. The great volume of air traffic over the Atlantic can be aided, then, with closer spacing of airplanes with increased safety.

So you see, the educational experiment you have just participated in is part of a great breakthrough in satellite communications.

In fact, under a long-standing agreement, a year from now our Government will loan the satellite to India. ATS-6 will be moved to a point over the Indian Ocean, where for one year it will be a vital tool in education, beaming instruction to your fellow teachers and their students in India.

The Indian Government has already built a ground transmitting station in Ahmedabad, near Bombay, from where the program will be beamed to the stationary satellite, over 20,000 miles up in the sky. The satellite will beam the signal back into the Indian subcontinent, where 5,000 TV receivers will show the program in 5,000 classrooms. If this large-scale experiment proves successful, as everybody expects, the Indian Government plans to go all out and put the entire Indian School System on satellites. They consider it the only feasible way to break the back of illiteracy which for centuries has impaired India's economic development more than any other cause.

Space developments, often derided as a reckless waste which only the super rich can afford, are thus beginning to provide the only effective solution to enable the poorest among our fellow men to help themselves.

Other countries are planning domestic communications satellites. Besides the United States, Canada and the U.S.S.R. already have such systems in operation.

Regional systems are under consideration by groups of nations. In this country, the Federal Communications Commission has officially opened up the domestic satellite communications market to private companies. As a result, we are witnessing quite a scramble of competing companies to get into the communication satellite business. One of the leaders in this new venture is the American Satellite Corporation, which recently began transcontinental satellite service, offering customers lower rates than they can get by using conventional overland telephone lines. American Satellite Corporation is a subsidiary of my company, Fairchild Industries, which, of course, built the ATS-6 satellite you have been using.

I see a tremendous future for these communications satellites. They will not only relay television network programming and carry telephone calls, they will also connect computers for nationwide operations ranging from airline ticketing to distributor support and inventory control. They may even have a beneficial effect on our environment by reducing the necessity for much business travel, thus contributing to reduced jetliner air pollution and conservation of petroleum resources. Communications satellites will make electrons do the traveling, with businessmen able to attend conferences from their own offices via closed circuit satellite television.

For the rest of the seventies and the early 1980's then, we will see a maturing and extension of the present satellite systems, the wholesale introduction of domestic satellite systems with capabilities tailored for each nation's needs, and finally the dissemination of television and radio signals directly from satellites to home receivers.

At present, when you sit at home watching television and you see the words "via satellite," it means the program or picture was relayed at some point by a satellite back to a ground station where your local TV station broadcast it in the conventional manner. The ability to broadcast television signals directly from satellites to home receivers will come as satellite power increases—even beyond the power of ATS-6—which now needs only a large antenna.

Satellite power and channel capacity have been limited by the cost of putting each pound of payload into orbit using expendable booster rockets. The cost is somewhat in the neighborhood of $10 thousand per pound. However, just around the corner is the space shuttle, a rocket that can be used over and over again. It will be launched like a conventional rocket, perform its mission in orbit, and then land much like a conventional jetliner. The reusable rocket will bring down the cost of communications satellites. The shuttle will also allow space crews to visit and repair existing communications satellites, or even assemble an entirely new satellite in orbit. Electricity generating solar cells from older, retired satellites, for example, could be assembled on a new communications satellite to provide super power for direct satellite to home TV set broadcasts.

The international political ramifications of this new communications potential are extremely challenging. The nations of the world will have to settle the question of whether one country has the right to broadcast satellite television programs to the citizens of another country.

A hundred years from now people will look back and wonder how man ever managed his affairs on his limited planetary abode without the tools provided by the space program. That there could ever have been a world without spacecraft will be just as difficult for them to perceive as for us one without telephones and airliners.

In 2073 A.D. mankind will be faced with many problems of a global nature that can only be tackled successfully by a global, coordinated approach. Successful solutions to these problems will require continuous monitoring and surveying of the entire Earth, supported by an effective communications system capable of transmitting the large quantities of collected data and picture information in real time to a multitude of users in nearly 100 nations. Unmanned Earth survey and communications satellites and orbiting manned observatories will form the backbone of this system.

The dominating problems of humanity during the next 100 years that require such aid from space will result from the collision between man's happy-go-lucky concept of unrestrained growth—both in numbers of people and their material expectations—and the grim limitations of resources the Earth can provide and of wastes it can absorb.

I ask you to visualize the world after the communications explosion: where disaster warning is a universal reality; where the finest library is available (via satellite facsimile) to the remotest outpost; where telemedicine brings the benefits of expert consultation to the rural doctor; where Nobel prize lecturers speak to high school classes; where ships at sea and aircraft in flight have quality telephone service.

Human language evolved because man felt the need to communicate non-real-time information and abstract ideas. Perhaps as a result of the communications explosion mankind can evolve a universal language. Certainly, space development—once derided as a waste which only the richest nations can afford—is beginning to provide the only solution to one of the most difficult problems facing all nations, rich and poor alike—effective communications.

*** END ***

– Chapter 46 –

American Ceramic Society Annual Meeting
Washington, D.C., May 9, 1972

Uses of Space and Space Technology [102]

WE ARE NOW IN THE 15th YEAR OF OUR NATIONAL SPACE PROGRAM. IT IS PERHAPS the most publicized non-military government activity in our history. Certainly it is among the most open to the public.

Millions of words describing space events have been published in the press over the past dozen years. Hundreds of millions of people all over the world have watched our astronauts carry out a spectacular exploration of the Moon on five different occasions. In NASA we work in a goldfish bowl, for all the world to see.

And yet, the space program is one of the least understood, and most misunderstood, activities undertaken by the U.S. government, to which the public, the press and the TV networks have had such free access. It is a curious fact that many foreigners, especially in the undeveloped countries, understand the importance of the space program better than many Americans. This is due, it seems, to a much clearer understanding of the value of science and technology in raising them from a grinding poverty to an easier, more rewarding life. Where some Americans, taking their very advanced technology for granted, see only its abuses, people in undeveloped and less technologically favored nations envy us our extraordinary standard of living and affluence. They rightly link an advanced technology with an advanced standard of living.

Today we live in a different world because the United States, in 1958, recognized the challenge of space and boldly made the necessary national investment to meet it. Since that time, more than a billion children, Earth's first space age generation, have been born in all parts of the world.

Because of space exploration, these children are learning a new cosmology, a new science, and a new view of man and his destiny in the universe. Where we of the older generations are still shackled by age-old concepts and conditions under which mankind has developed for more than a million years, today's children can look forward to the opening of entirely fresh opportunities for enterprise and intellectual and spiritual growth. For, make no mistake, the most profound impact of the opening of space to man will be on the mind and spirit.

We are all aware of the mounting expenditures of the federal government. Most of us would like to see either a leveling off, or an increased ability to pay for national programs considered vital to American life. A plateau for the federal budget just does not seem to be in the cards. There are far too many demanding goals to be achieved, goals involving social and environmental improvement, which are very costly. So we must take the route of increasing our resources to pay for these social, urban and environmental improvement programs. Unless this is done, there will indeed be a drastic reordering of our national priorities and a leveling off of the very programs for social betterment we wish to expand. Even more important, unless we increase our resources and technological capabilities, we shall become less and less viable as a nation capable of meeting the needs of its people. In that direction lies only disaster.

In the United States, we have been spending more and more of our tax dollars to effect social change: in our schools, and in our health, welfare and poverty programs. These programs are, in reality, a redistribution of available resources. Our returns on this investment are the satisfaction and enjoyment of a better society, and a greater fulfillment of our desire for a higher civilized state. To carry out these social programs is a prime function and responsibility of society through its government.

We have also been spending our tax dollars in NASA's aerospace programs, but these programs, unlike those in social welfare and improvement, create new resources for the United States—and, indeed, for the entire world. Our investment, in this case, pays back in the form of a stronger technological base, a more advanced science, and a more comprehensive understanding of our Earth, the solar system, and the universe of which we are a part. In addition, we are already realizing dollar returns from our activities in space, and from space technology which was not anticipated at so early a time some 12 or 14 years ago.

Now, you might wonder whether having a strong technological base, a more advanced science, or a more comprehensive understanding of this planet and the rest of the universe is so terribly important. Aren't they rather secondary to our immediate problems?

Let me begin by reminding you that our standard of living—any standard of living—is firmly rooted in technology. Without our machine tool technology, our electric energy-producing technology, our communications and transportation technologies and, yes, our ceramic technology, and the thousands upon thousands of other technologies on which civilization is built, we'd be back in the Stone Age where the prime goal was to stay alive.

So a strong technological base is a necessity to modern civilization. But you can't build a technological foundation to a certain level and then forget it. It must be continually updated by developing new, more advanced technologies, because society itself doesn't stand still. Even without population increases, societies become more sophisticated in life concepts and desirable goals to be achieved, giving rise to requirements the old technology cannot adequately meet. Economic and social stagnation are the result of a prolonged stagnation of technological development.

Well, what is the role of science, then? In science we explore the nature of things, of phenomena of the atom, of life, the Earth, of space, the solar system and universe, of man. Science seeks to identify the principles of creation, not just to satisfy our thirst for knowledge, but to instruct us about ourselves, to understand the dynamics of our environment, to comprehend ever more of the cosmos to the end that we may move more intelligently toward a more enlightened destiny.

But science is also necessary to provide the ideas and knowledge on which new technology can be developed. Science broadens our vision so that we can see how to fulfill further human requirements. But it also provides the basic principles on which to build the technology to effect those requirements.

So a dynamic, growing civilized society must continually add to its scientific knowledge for both cultural and practical reasons. A nation or people that does not press forward on the frontiers of science must in the end stagnate and regress, its spirit and imagination debilitated.

Lastly, we are literally driven to comprehend our environment, our planet and its environment, the solar system, and the universe, in order to survive. Man has reached the point in his evolution where he is consciously altering his own ecology and the ecologies of all living things. For centuries we have been doing this unconsciously, but now it is no longer a blind act. To stop now is impossible. But a comprehensive understanding of our environment and the dynamics governing it will lead us away from a disastrous, uninformed meddling to a more constructive conservation of life and resources of the planet.

Now, where does the space program fit in all this?

The answer is, right in the middle, and all through it. The space program is a prime factor in providing us with a strong technological base, with a more advanced science, and with an increasing understanding of the Earth, the solar system and the universe, all of which are essential if our goal of greater quality in life is to be realized.

We use the techniques of space flight to increase our knowledge of the laws governing the universe by detecting, measuring and analyzing the phenomena of space, much of which we cannot study from the Earth's surface. The spacecraft that we place in Earth orbit to study the stars, and others that we send to explore the planets, or the astronauts who explore the Moon, are examples of the things we do to increase our knowledge of the great cosmos of which we are a part. Space technology therefore is important for a very basic reason: scientists can place their instruments in locations which previously were inaccessible to them. Beginning in the next decade, they will be able to use the space shuttle to carry both themselves and their laboratories into near-Earth space for more extensive experiments and observations than can be conducted from the ground.

So space astronomy is important because it is the only way in which we can observe and obtain data of phenomena and events which affect our lives here on Earth. For example, one of our instruments—a coronagraph—last year revealed three luminous jets of intensely hot solar plasma being hurled out into space at a speed of two and a third million miles per hour from the Sun. These hot jets were 20 times the size of the Earth, and represented the energy of 100 million atomic bombs. A peak temperature of 30 million degrees must have been attained in this event, producing matter in a physical state man cannot produce on this scale on Earth, and can only observe in astronomical sources.

Study of our Sun is, of course, important because it controls the weather and climate of Earth as well as sustaining all life. But also, by studying its sources of energy and other phenomena, we may learn how to produce and control new sources of energy we shall require in the future.

Many of us have been intrigued and excited by the views of Mars which we have received from our Mariner spacecraft now orbiting that planet, or of the mysterious, cloud-shrouded planet Venus revealed by other Mariners. But, from a practical point of view, we might ask, how does the exploration of Mars and Venus directly benefit mankind?

The fact is, we are gaining important data from measurements of the atmospheres of these planets which could eventually help us determine the basic causes governing the origins and motions of the complex weather systems we have on Earth. Three factors make the dynamics of our atmosphere very complicated to study. These are rotation of the Earth, the oceans of water and water vapor in the atmosphere, and the uneven heating of the Earth's surface because of the particular distribution of land and seas over the globe. Thus, the long-range prediction of weather becomes extremely difficult because it is impossible at this time to isolate the effect of each of these factors.

Now, however, by studying the atmospheres of Mars and Venus, we can begin to evaluate the importance of each of these parameters which affects the circulation of the Earth's atmosphere. This is because Mars' atmosphere is thin, there is hardly any water vapor present, and no oceans on the surface. Therefore, the Martian atmosphere is in simple radioactive equilibrium and its motions respond to direct solar heating, and are further controlled by topography and rotation.

Venus, on the other hand, has a very thick atmosphere, a dense cloud cover, and rotates very slowly. Thus radiation is unimportant and the temperature structure is controlled by variable wind systems.

By studying these "simpler" atmospheric systems on Mars and Venus, we hope to provide some of the needed insight into the basic causes which generate the weather systems on Earth.

But the uses of space and space technology go beyond providing us with a better astronomy and planetology, important as these studies are in comprehending the solar system and universe for both cultural and practical reasons. There are more direct applications in such things as communications and weather satellites, with which most of us are familiar. But perhaps not everyone is familiar with the increased productivity of our latest *Intelsat IV*, which provided 9,000 equivalent voice channels compared with 240 on *Intelsat I*. This has meant a two-to-one reduction in the cost of leased lines across the Atlantic, and a two-to-one reduction in overseas telephone rates.

Our next generation Tiros weather satellite will far exceed *Tiros I* in the collection of data, and our ultimate goal is accurate two-week weather forecasts.

But our most exciting space vehicle system will soon begin to study the Earth. This is the Earth Resources Technology Satellite.

With ERTS, we shall begin to learn how to monitor our environment from several hundred miles above the surface. Among valuable data ERTS is designed to obtain are included the amount of water runoff to be expected from the observed snow coverage; the quantity, quality and distribution of crops, and when to expect the harvests; how to determine ocean resources; and relate geologic structures to mineral resources. ERTS also will provide up-to-date maps of present land use, which will help in planning future land development. Forestry services will benefit from the monitoring of forest resources, including the state of insect infestations, disease and kinds of trees. Programs to preserve the environment will be aided by the monitoring of sources of pollution in waterways and the air.

The ERTS program is to be aided by our *Skylab* experimental space station to be launched next year. The astronauts will supplement airborne and satellite observations with their own Earth resources experiment package. They will use, among other experiments, multispectral photography, consisting of a battery of six cameras that simultaneously photograph the same area, each viewing in a different wavelength. By this means, various conditions of interest can be observed, such as soil moisture, types of vegetation, plant health, and so on.

An important function of the *Skylab* experiments is to compare results with concurrent measurements and observations from aircraft and on the ground. This will aid in the development of the necessary instruments and equipment to be used in satellites sooner than would be possible without human monitoring of their test operations in space flight.

We see a great deal of promise in ERTS. Although some of our expectations may not be realized, yet I feel the real benefits of environmental and ecological surveys from space will surprise many, even those of us close to the project.

Few people have appreciated the way in which the space program pulls into its train the activities of a vast number of scientific and technological disciplines, stimulating research and innovation in fields that at first glance would seem to have no connection with space flight. For example, NASA has been making contributions to knowledge of the basic building blocks of life. This work adds to our understanding of just what life is, and the reason for our interest is simple: we want to be able to recognize life should it exist on other planets or, perhaps, in space itself.

Some of NASA's studies are less direct in effecting advances in other fields. One of our scientists, for example, a specialist in space radiation effects on the blockage of cell division, has devised and demonstrated a theory that helps explain cancerous growth. His theory also indicates the development of chemical countermeasures against cancer. Without going into too great detail, it seems that cell division, according to the theory, is controlled precisely by the pattern of ion concentrations on the surface tissues of the cells. The pattern is formed by the electrical voltage that normally exists across the cellular surfaces and varies from one part of the body to another.

The scientist noticed that cells having a large negative electrical potential seldom if ever divide. On the other hand, cells with small negative potential divide at maximum rate. Tests showed that ion concentration differences between membranes did indeed exert a powerful control over cell division. Thus, the implication of this research is that the primary change which occurs when a normal cell becomes malignant consists of a basic functional change in the molecular structure and special characteristics of the cell surface.

The point is that this particular finding might never have been discovered except for the investigations into cosmic radiation effects on cells.

The ladies present here who are concerned with providing their families with safe, nutritious food may have special interest in NASA's program in developing food technology. As might be expected, we must have extremely rigid requirements for the food to be used by the astronauts, since any type of food system failure could have very grave consequences. It must be free of contamination, be of a high, known nutritional value, and not require refrigeration, even under wide temperature variations for long periods of time. The precautions we have taken with food for the astronauts have led to new and improved methods of processing, preserving and sterilizing it. NASA has been responsible for development of nearly 100 different freeze-dried or dehydrated foods. One NASA contractor has developed a method of keeping bread in storage for more than 14 weeks with no sign of mold growth. Another has developed "instant" rice which can be served in three minutes from shelf to table. Still another has come up with precooked, pre-buttered rolls which can be preserved for up to 600 days.

Food technology has advanced considerably from the early days of manned space flight. We have been able to add appeal and acceptability to our requirements with some success. Our methods assure reliable quality, and the famous "clean room" conditions developed for certain spacecraft and equipment construction are used in the processing of food. Using NASA-developed methods, certain foods can be prepared and stored for emergency situations. When a disaster occurs, the food could be immediately shipped to the stricken area with no loss in preparation time and no need for refrigeration, even in warm countries.

The fact that more than 60 percent of all food consumed in the United States is significantly preprocessed or prepacked indicates that any advances in this field are likely to have wide and beneficial effects. The increasing problems the food industry is having with food additives, nutritional quality and preservation make it seem that NASA's work in developing good, safe food for the astronauts will provide considerable assistance in improving the food eaten by Americans.

There isn't time to describe the many different fields which are being aided by space-developed technology. But they include clothing, paints, machine tools, shipbuilding techniques, oil prospecting, ship and airline navigation, medical x-ray diagnostics and remote monitoring of hospital patients, among thousands of others. However, I don't want you to think the space program is being justified because of these "spin-offs" from our activities. Our main business is to explore and discover, develop new national capabilities of space flight and uses of space, to increase our scientific knowledge and develop the kind of high technology this nation requires in order to survive in a highly competitive world.

NASA is a creative organization. The investment of 1.3 cents on the tax dollar is returned in the form of a stronger America economically and a broader range of technological capabilities which offer the nation many options for future greatness. The space shuttle which is now being developed will place in America's hands a vehicle which will transform space travel from a Buck Rogers epic into a routine transportation vehicle to be used eventually by average people. Space is the great future of the young generations. It is they who will find the unlimited uses of space of which we can only imagine.

*** END ***

– Chapter 47 –

"Almost a confessional," a colleague described von Braun's last major piece of solo writing. Sometime in 1976, he accepted an invitation from the Lutheran Church of America to present a major paper at its synod at the University of Pennsylvania that autumn. He worked on the paper for several months in the hospital. When the time came for its presentation, he was too ill to appear. He had a surrogate step in and read it to the large assembly.[103] Von Braun was diagnosed and treated for cancer shortly after his move to Washington, D.C. in 1970.

Almost the whole world had witnessed the gigantic strides of von Braun's life and work. He relished the incredible popularity of his space achievements. The final phase of his fabulous life in the sick room of Alexandria Hospital was marked by seclusion and silence. *"Thy will be done"* was his last credo. Wernher von Braun ended his Earthly voyage on June 16, 1977.[104]

Responsible Scientific Investigation and Application [105]
A Talk Presented to the Lutheran Church of America
Philadelphia, Pa. 29 October 1976

LADIES AND GENTLEMEN:

When President Marshall invited me to this symposium, and suggested that I present my thoughts on responsible scientific investigation and application, I wondered whether I was really up to this task. Science, after all, has many facets and I have had only a limited view of a very few of them. And scientific applications are being made in the entire spectrum of human activities.

Moreover, the word "responsible," when applied to scientific investigation and application, has several widely different connotations. For instance, is it financially responsible to support a costly new science project whose payoff in saving lives, gaining new basic knowledge, or leading to new consumer products may be difficult to assess? Or is its support ethically responsible as we weigh the good and the evil capabilities that may spring from it? Is the support of a new science project professionally responsible in view of a possible controversy among the investigators and scientific peers about the validity of its scientific objectives?

After some reluctance I decided to take the plunge and try to present you with some of my views on a number of subjects that I felt were at least relevant to the overall topic—without attempting to be responsive to all its aspects. The subjects I selected are:

Motivation for Scientific Study and Research.
Science, Technology, Morality and the Question of Taboos.
Mankind's Most Pressing Problems.
A hierarchy of Scientific Priorities

And finally (as this is a church-sponsored symposium) some

Thoughts About Science and Religion.

If some of you may find my paper here and there a little heavy on the philosophical side, please consider this as a groping search for the core of the question. Your theme, "Responsible Scientific Investigation and Application" poses undoubtedly one of the most fateful problems of the 20th century. It cannot be done justice with lofty generalities or simplistic pontifications of what are often extremely complex and heart-renting problems. I found it necessary time and again to quote specific cases and propose possible solutions for them. And it is at this point that I must ask your forgiveness in advance. You will find that I selected most of these examples from the field I am most familiar with, namely space science and technology. I did this merely because I felt both you and I would be more comfortable if I applied my conclusions to a field where I am at home. Please do not construe this as an attempt to bias our scientific priorities in favor of space. While I do believe with all my heart that our space program has made, and will continue to make, great contributions in many disciplines of science and their practical applications, I am fully aware of the pressing needs in many fields completely unrelated to space.

1. Motivations for Scientific Study and Research

Ralph Waldo Emerson once said, "Men love to wonder, and that is the seed of science."

Wonder, curiosity is indeed the mainspring of all scientific research. Since time immemorial, there have always been some men and women who felt a burning desire to know what was under the rock, beyond the hills, or on the far side of the open seas. With all the rocks and hills and oceans explored, this restless breed now wants to know the nature of subatomic particles, the mechanism through which given forms of life can make duplicates of themselves, whether there is life on Mars, or whether the whole cosmos started with a Big Bang.

Why does man want to know all these things? Why are we so eager to learn more about the universe in which we live?

I cannot provide you a better answer than a quote from that great explorer and humanitarian, Friedjof Nansen: "The history of the human race is a continual struggle from darkness toward light. It is, therefore, of no purpose to discuss the issue of knowledge. Man wants to know and when he ceases to do so, he is no longer man."

Even from the purely materialistic angle, man's pursuit of his innate curiosity has stood him in good stead. All material advances so enjoyed today, whether color television or miracle drugs, jetliners or aqualungs, sewing machines or artificial kidneys, can be traced back to the fact that at some time, somewhere, someone was curious about something.

As a man whose entire life has been devoted to rocketry and space exploration, I should like to make a few remarks about the relationship between space research, scientific education and motivation of the young who we expect to continue where my generation leaves off.

The space program really did not begin with *Sputnik*, *Explorer I* and first manned flights into orbit. Our program of space exploration is simply the latest chapter in a continuing scientific revolution that dates back to Copernicus, Kepler, and Galileo. It was their observations, and their ability to draw inferences from what they say, that moved the Earth from its exalted position as the center of the universe. That was a profound blow to the human ego, and it called for some very painful reassessments of man's place in the universe. But it was a very necessary step in the continuing search for truth and the advancement of knowledge.

Today every school child is aware that our small planet, its neighbors in the solar system, and even our Sun itself, are in what someone has described as the suburb of a minor galaxy in a dynamic universe populated by galaxies and super galaxies in numbers that probably surpass our comprehension.

The average citizen today, of course, has far more scientific information at his disposal than did those greatest of intellects of earlier times. Yet paradoxically, I think there has never been a greater need for increased understanding and appreciation of science and technology.

The noted educator, Dr. Lindley J. Stiles of the University of Wisconsin, addressed himself to this question of scientific literacy some years ago. Dr. Stiles said that, although the choice of direction for our civilization will be determined through the democratic process, it is there that the problem begins. To make rational choices, he pointed out, the average citizen must understand the nature and role of science at a time when its breadth and complexity are increasing almost exponentially.

Conversely, the scientist, at a time when he can barely keep up to date in his specialty, must not isolate himself in his parochial interests. Instead, he should see his profession as a part of the larger world, to evaluate himself and his work in relation to all forces, especially the humanities, which shape and advance society.

The need, then, is for an educational process resulting in more scientific literacy for the layman, and more literacy in the humanities for the scientists. It is also important that the layman not attach too much importance to the scientists' opinions on issues outside their special disciplines. Scientists are not experts in everything just because they are scientists.

Man in this scientific and technological age is free only to the extent he has a grasp on himself and his surroundings. Freedom—the ability to speak, think, act and vote intelligently—is based largely on our ability to make choices growing out of our understanding of the issues involved. With each advance of science, and with each invention of technology and its uses, there is an invitation to more understanding. This is the essence of the burden borne by all peoples since the dawn of humanity and tool-making. This is the imperative for scientific literacy and, we should add, technological literacy. There must be widespread understanding of the role of science and technology in modern society, both as to their limits and our dependence on their basic function as tools for our survival.

How do we encourage scientific and technological literacy? I think the problem is how to instill in students a permanent desire to learn.

All youth is endowed with curiosity from the very beginning. What can the education process do, not only to keep this natural curiosity alive, but to make it a permanent part of the individual drive?

Professor Okey of Indiana University offers one approach: "In addition to learning facts," he wrote, "students should learn to examine facts, how to answer questions or solve problems using facts, and how to produce facts."

This is essentially the scientific method. By learning the scientific method, students will understand its role in society and at the same time learn to think for themselves. Learning to think for oneself, in turn imparts a deep sense of freedom. Once tasted, an appetite for it is formed which may well endure throughout life.

But if our young people are going to gain this appetite, our schools, our colleges, our universities, must bear an ever greater responsibility. All too many times in the past, education—particularly in the scientific disciplines—has placed extremely heavy emphasis on transmitting the established knowledge of the past. There has been a tendency for teachers to assign reading, and to encourage rote learning, instead of taking the admittedly more difficult path of encouraging students to think for themselves.

A serious trend that I see emerging is the intellectual effort to bridge the gap between the natural sciences and the social sciences. This effort must by its very nature involve the educational processes at all levels. We see it manifested today in the pronouncements of many of the professional associations of science, as well as of prestigious natural scientists. We see less of a realization of its importance among the social scientists and humanists. But here, too, the realization is beginning to grow in the writings and pronouncements of such prophets as Buckminster Fuller and the poet MacLeish. In my opinion, democratic society cannot survive if the masses of the population are lacking in understanding of the fundamental principles of science; it is through science and its related technology that more and more of the operations of society and the decisions that will have to be made concerning these operations are based. From an educational point of view, this effort to bridge the gap involves, as I see it, reorientation of curricula in both the natural sciences and the social sciences. This, of course, takes time and thought, because the curricula in each of these fields have tended to go more and more their independent ways.

Never has there been a greater opportunity, or a greater incentive, for the young people to learn to think for themselves. The body of knowledge is advancing at an absolutely incredible rate, and new tools are constantly being made available to the researchers that were previously not only unavailable, but often, not even dreamed of.

Without wanting to seem overly partisan, I would like simply to point out that the space program has, by all standards, become America's greatest generator of new ideas in science and technology. It is essentially an organization for opening new frontiers, physically and intellectually. Today we live in a different world because in 1958 Americans accepted the challenge of space and made the required national investment to meet it.

Young people today are learning a new science, but even more important, they are viewing the Earth and Man's relationship to it quite differently—and I think perhaps more humanly—than we did fifteen years ago. The space program is the first large scientific and technological activity in history that offers to bring the people of all nations together instead of setting them further apart.

2. Science, Technology, Morality and the Question of Taboos

One of the most disconcerting issues of our time lies in the fact that modern science, along with miracle drugs and communications satellites, has also produced nuclear bombs. What makes it even worse, science has utterly failed to provide an answer on how to cope with them. As a result, science and scientists have often been blamed for the desperate dilemma in which mankind finds itself today.

Science, all by itself, has no moral dimension. The same poison-containing drug which cures when taken in small doses, may kill when taken in excess. The same nuclear chain reaction that produces badly needed electrical energy when harnessed in a reactor, may kill thousands when abruptly released in an atomic bomb. Thus it does not make sense to ask a biochemist or a nuclear physicist whether his research in the field of toxic substances or nuclear processes is good or bad for mankind. In most cases the scientist will be fully aware of the possibility of an abuse of his discoveries but, aside from his innate scientific curiosity, he will be motivated by a deep-seated hope and belief that something of value for his fellow man may emerge from his labors.

The same applies to technology, through which most advances in the natural sciences are put to practical use. No sooner had man learned to make iron, he beat it into axes to fell trees so he could till more land, and into swords to defend himself or grab land from his neighbors. The modern offspring of that axe, the knife, may save a life when yielded by a skillful surgeon, but will kill if thrust only a few inches deeper. Aircraft carry millions of people across oceans and continents, thus forging ever-closer ties of understanding, friendship and trade between nations, but are just as capable of carrying troops, war material or even nuclear bombs. Technology, all by itself, thus does not have a moral dimension, either.

The deplorable fact that neither science nor technology have a moral dimension, of course, does not help solve the problem we are trying to address here today.

Science and technology have made such rapid advances in so many fields that the public is almost ready to expect they can solve any of mankind's problems if we only set our mind to it and make the necessary resources available. "If we can send men to the Moon," many people seem to feel, "why can't we— " and then we hear an endless enumeration of all the grim problems besetting humanity.

Many of mankind's problems are, of course, outside the domain of science and technology. but probably just as many are profoundly affected by them. And we have reached the point in many fields of scientific and technological endeavor where we can expect success of a new project with a very high degree of probability, provided we are willing to spend an adequate amount of effort and money. The decision whether a proposed new objective should be pursued must, therefore, often be made not on the basis of whether it is attainable or not, but on its general desirability, its intrinsic value, its future growth potential, its inherent dangers, its environmental impact, or its cost compared to other competing science projects.

The choice must also be based on a new value system. The old American standards of material wealth or technological efficiency obviously require some reappraisal, since in these terms we are already the greatest society in history. As Professor Herbert J. Muller, professor of English and government at Cal Tech once formulated it, we should adopt "some civilized standard, involving moral, cultural, spiritual value, the kinds of achievements recognized in the broad agreements upon what were the great societies and the golden ages of the past,"

Also, we are squarely confronted with the ethical question: Is everything that is scientifically possible, also permissible?

There can be no question that unless we find a more effective mechanism to steer scientific investigations and their practical applications into constructive and safe channels, the future of mankind will be exposed to ever-growing dangers. As we have learned to harness and unleash fantastic new sources of energy, our potential to create a life of abundance for everybody and to bring about a sudden apocalyptic end has grown immensely. In the horse and buggy days, no one was exposed to great risk if the coachman had a drink too many. Today, the driver of a high-powered automobile can become a killer in one careless moment. Tighter rules of the road have become a necessity for survival.

But how do we write a set of rules of the road for scientific investigations? Should we attempt to develop a set of taboos declaring certain scientific studies off limits? If so, who should be entrusted with the task of setting those taboos?

Looking back to mankind's slow emergence from caveman to spaceman, from cannibal to cardinal, we may well doubt whether we would have ever succeeded, had such taboos been strictly enforced in the past. Take the case of Prometheus. Clearly, had he not defied the Olympic taboo we would still be shivering in unheated caves. And yet, suppose Zeus in all his wisdom would have convened an Olympic committee, complete with futurologists and environmental impact appraisers, to advise him on the foreseeable human sufferings that had to be expected from the fires running out of control, or by the use of fire in war. Is it not likely that that committee would have produced powerful reasons supporting Zeus' final decision to pronounce the Prometheus taboo? And in retrospect, are we not fortunate indeed, that Prometheus defied the ruling?

How much smarter are we today? Could we really convene a group of academic Olympians wise enough to set the green and red traffic lights for fundamental research in such equally promising and dangerous areas as genetic engineering, breeder reactors, high power lasers or the harnessing of thermonuclear power?

Dire predictions have been made that even the *next* generation must gird itself to cope with near hopeless problems resulting from the divergence between the growing demands of a rapidly expanding world population and the shrinking supply of non-renewable resources of our planet.

Is this really the time to put shackles on scientific research in areas that are both highly promising and fraught with grave hazards? Is this the time to put Prometheus in chains before he can carry the new fire to man? Don't we need all the new insights that only unfettered scientific research can give us if we want mankind to survive, not just the next generation, but for a few more million years, and hopefully the remaining four billion years the Earth is expected to remain inhabitable by man, unless he himself deprives him of that chance.

And yet, has mankind really a chance to survive the remaining several millions or even billion years—or even just the next generation—if we refrained from imposing any effective controls? I think the majority of scientists are convinced that very wise and cautious statesmanship and ironclad international treaties are absolutely mandatory to protect mankind from the abuse of its newly-won scientific capabilities.

The question, therefore, boils down to the best point in time between gestation, birth, childhood, adolescence and maturity of a new potentially dangerous discovery or invention at which these controls should be imposed.

It appears to me that the time to flash up the red no-go sign is *not* at the very early stage of any basic scientific pursuit. As Plato said: "We can easily forgive a child who is afraid of the dark; the real tragedy of life is when men are afraid of the light."

Controls should rather be established only after the area of potentially dangerous abuse of an otherwise promising and beneficial scientific discovery can be clearly identified and isolated. No judge would condemn a human embryo to death because it has the potential to become a criminal. Laws of dos and don'ts are rather applied to children and grownups with clearly identified legal and illegal options, and the law is enforced by policing, and imposition of penalties in case of violations.

I fervently believe that not everything that is scientifically possible is *per se* permissible, in the ethical or religious sense. But the stop light for scientific research and application should be applied only after we can discern the beneficial from the detrimental, the good from the evil.

Do we need a new organizational structure to administer whatever new controls in the field of scientific investigation and application may become desirable? I do not think so. In the medical and health fields the National Institute of Health, the National Cancer Institute and the National Food and Drug Administration are examples of organizations already deeply involved. In the food production area the Department of Agriculture must continue to work out its differences over fertilizer and pesticide uses with the Environmental Protection Agency. And in the question of nuclear safety for our many new nuclear power plants we simply cannot find any more competent people than in the Energy Research and Development Agency and the Nuclear Regulatory Commission.

Wherever fundamental new research is required to provide a better base for future legislation—for instance in the field of absorption of pollutant and toxic metals by the whole complex food chain of marine organisms living in our river estuaries and on our continental shelves—the National Science Academy, the National Science Foundation and the National Oceanic and Atmospheric Agency are competent, eager and busy formulating the necessary programs.

3. Mankind's Most Pressing Problems

The most pressing problems of our generation are easily identified. They are probably subject to a little fundamental disagreement among most Americans, and many apply to mankind as a whole. None of them can be solved by science and technology alone. All have economical, social, philosophical, aesthetical, political and other aspects which are often in conflict with one another.

My paper deals with responsible scientific investigation and application, and attempts to identify such problem areas where science can really help. For this reason I have intentionally omitted problem areas on which science policy has no profound effect, such as Africa's new search for identity, the painful evolution of democracy in some Latin American countries, or the immensely complex Middle East problem with all its ramifications. I excluded these problem areas not because I underestimate their possible fateful significance, but solely to bring my science-related message into better focus.

A dome supported by five pillars offers a maybe somewhat simplified model for the hierarchy of our key problems. The dome itself represents the all embracing overall objective of "Survival of Mankind as a Species." The five pillars supporting the dome are:

1. Resources survival: here we are talking about the continuing availability of the material and energy resources required to feed, clothe, house and provide jobs for the world's growing population.
2. Environmental survival: we must drastically clean up and protect from further deterioration the precious life support system of our thin biosphere, the thin layer of soil, water and air that supports all life on Earth. This will require continuous monitoring of the land and ocean surface and its atmosphere.
3. Spiritual survival: we must accomplish all this and more without drifting into a regimented, coercive society. To restore and protect the inalienable personal freedom we must provide more elbow room for the individual. This requires a reversal of the trend toward urbanization.
4. Nuclear survival, which is tantamount with avoidance of a self-inflicted nuclear holocaust.
5. Scientific survival: Unless the United States retains leadership in the natural sciences and in technology, these four support pillars for the dome we called "Survival of Mankind as a Species" are bound to erode and collapse. Scientific survival, therefore, is a vitally needed fifth pillar. We live in a dynamic, fast-moving world. The great races for superiority in aviation, nuclear and space technology between the superpowers have dramatically illustrated that he who relaxes his efforts in science and technology, even for a few years, can be easily confronted with a momentous, fateful and possibly hopeless task of catching up again. But in continuing our pursuits in applied science we must never neglect scientific effort for its own sake, for the continued search for the basic laws of creation. It is basic research which lays the groundwork for any scientific application.

A Hierarchy of Scientific Priorities

Using our dome-and-five-pillars model, we can now attempt to identify some of the most promising contributions of applied science and technology. Let us look at each pillar and see how science and technology can protect and strengthen it. Needless to say, science and technology have so many facets that my list of scientific priority projects is incomplete and should be extended to include other relevant programs aimed at protecting our five pillars.

4.1 Resources Survival

It has been said that the Earth is a spaceship with 3½ billion astronauts and no captain, coasting through the universe to an unknown destination. And while they are rapidly depleting the ship's limited supplies, those astronauts are multiplying like rabbits.

Well, regardless of where you stand on that unqualified demand for "limits to growth" expressed in some quarters, and the dire predictions of the "Club of Rome"—which very nearly comes to the conclusion that we are doomed no matter what we do—it should be obvious that a systematic survey of the Earth's resources should have one of the highest priorities for science and technology. It is in this field, where space technology has provided us with a most timely breakthrough.

In Space parlance, we call "resources" everything man needs for his survival: food crops, timber to build houses, fresh water, wildlife and marine life to hunt and fish, cattle and grazing land, minerals, fossil fuel and other energy sources (such as solar energy and uranium), and last but not least, remaining land for his future use.

Two NASA-built Landsat satellites, orbiting the slowly about the revolving Earth 14 times a day in near-polar orbits, are swamping the data centers on the ground with a wealth of new information, and it will probably take quite a while until we will be able to completely interpret the maze of new orbital data on resources contained in their multispectral images. But we have already come down the road quite a bit. A customer anywhere on Earth buying images from both satellites gets a picture of the same place every ninth day. As of October 1975, more than 155,000 different pictures of parts of the Earth's surface were taken, each in four spectral bands, stored at, and available to the public from, a data center in South Dakota.

NASA trains foreign scientists to interpret the images; data goes to more than 100 research teams around the world, giving many less developed nations their first report on natural resources. Data have taken polluters to court; led geologists to oil; given land planners a regional picture faster, at far less cost than traditional surveys; analyzed plant health for farmers and foresters; corrected maps; showed the Sahel (African drought-famine regions) that some land had been preserved from encroaching desert by controlled grazing, etc. Such information from satellites has been called as epoch-making as the first use of fire as a tool, or the first practical use of the wheel.

There are, of course, many other areas where science and technology can make great contributions to mankind's growing resources dilemma. But there can be no doubt [that] continued aggressive research and development programs aimed at improved resources surveys from orbit is one of our best bets.

Let me now touch upon the vital subject of our continuing supply of energy. We must distinguish here between the energy problem for mankind as a whole and our national energy problem, which is overshadowed by the desire to minimize our dependence on imports which, as recent events have shown, make the United States vulnerable to undesirable pressures on U.S. foreign and domestic policies.

Let me quote from a recent publication from ERDA, our new Energy Research and Development and Administration.

> "The national energy system currently relies most on the least plentiful domestic energy resources and least on the most abundant resources.
>
> Over 75 percent of the Nation's energy consumption is based on petroleum and natural gas. Domestic supplies of these commodities are dwindling.
>
> Coal, the most abundant domestic fossil fuel, provides less than 20 percent of current energy needs.
>
> Uranium, the domestic energy source with the greatest energy potential, provides about 2 percent of the Nation's energy.
>
> Solar energy, available to all, but diffuse, provides a negligibly small percentage of current needs."

In response to this situation and after a careful analysis of projected needs, the status of candidate technologies and the extent of the resources they would use, a ranking list for ERDA's priorities has been developed and submitted to the Congress. The list breaks the priorities down into three categories, and here I am quoting again directly from ERDA's official plan:

"For the near-term (now to 1985) and beyond the priorities are:

> To preserve and expand major domestic energy systems; coal, light water reactors (the highest nuclear priority), and gas and oil from both new sources and from enhanced recovery techniques;
> To increase the efficiency of energy used in all sectors of the economy and to extract more energy from waste materials.

For the mid-term (1985-2000) and beyond the priorities are:

> To accelerate the development of new processes for production of synthetic fuel from coal and extraction of oil from shale;
> To increase the use of underused fuel forms, such as geothermal energy, solar energy for room heating and cooling, and extraction of more usable energy from waste heat. None of these technologies has a major long-term impact, but each can be quite useful in relieving mid-term shortages.

For the long-term (past 2000) priorities are:

> To pursue vigorously those candidate technologies which will permit the use of essentially inexhaustible resources;
> Nuclear breeders;
> Controlled thermonuclear fusion;
> Solar electric energy from a variety of technological options, including wind power, thermal and photovoltaic approaches, and use of ocean thermal gradients."

None of these three technologies is assured of large scale applications. All have unique unresolved questions in one or more areas: technical, economic, environmental or social. The benefits to be gained in achieving success in one or more of these approaches require that vigorous development efforts proceed now on all three.

In the context of my paper—Responsible Scientific Investigation and Application—I cannot take issue with any elements of this ERDA program. I think it is a mature and well thought out plan which, as all plans aimed at broad objectives, will require repeated reviews and adjustments to the ever-changing realities of costs, needs, encountered difficulties, accomplished scientific and technological breakthroughs and even national moods.

There is one item buried in the multitude of ERDA goals that requires particular attention, however. As we have seen, ERDA recommends for the near term a great expansion in the building of light water reactors which power the conventional nuclear utilities already providing a small percentage of our electrical power needs. For the long term, ERDA plans to pursue the breeder reactor program, so called because it produces more nuclear fuel than it consumes.

I am personally convinced that the nuclear safety aspects of both conventional and breeder type power plants can be met if the same standards of perfection and careful scrutiny are applied that made our lunar landing program successful. However, both types of reactors produce an ever-increasing amount of plutonium. The question of how to avoid an uncontrolled proliferation of this material which can be used to make atomic bombs is serious indeed. I shall address this question later in context of the Nuclear Survival pillar of our pillar-and-dome model.

Man's most precious resource undoubtedly is his brain, but the vast majority of mankind uses it most sparingly. The reason, of course, is inadequate education.

I have already talked a bit about scientific education, which we need to provide the world with new generations of scientists. Important as it is, the education of new scientists, of course, directly involves only a tiny minority of mankind. The sorry fact is that the bulk of the 3½ billion astronauts of spaceship Earth are illiterate. If we want to improve the quality of their lives, we must literally start with the ABCs.

In the summer of 1975, the United States, implementing a state treaty with the Government of India, placed a unique geosynchronous communications satellite above the African East Coast, from where it is sending educational television programs to 2,500 villages in India. The programs themselves—the "three R's" for the kids, and topics like farmer instruction, animal husbandry and family planning for the grownups—have been prepared by the Indians themselves and are beamed up to the satellite from a large transmitter station in Ahmedabad, near Bombay. The satellite, equipped with a huge 30-foot parabolic antenna and a powerful transmitter, beams the amplified signal back onto the entire Indian subcontinent, where it can be received in any village equipped with a normal TV receiver attached through a special black box to a 6-foot chicken-wire antenna. The entire TV village setup, a 20th century version of the little one-room red school house of early America, costs less than $1,000.

Both from the technical viewpoint, and from the aspect of audience acceptance, the Indian program has been a great success. However, we should not loose sight of the fact that it was only the first large-scale experiment of carrying education via a direct broadcast television satellite to a predominantly illiterate population. From this first experiment to an operational educational system for all of India is still a long, difficult road.

Just look at a few figures which illustrate the immensity of the problem. India is a very large, diverse country with an area of over one million square miles. It has 14 major languages and about 800 different dialects. Approximately 80 percent of its 500 million people live in rural areas. There are about 560,000 villages, many without access roads, electricity, hospitals, or schools. Seven out of ten people are illiterate. Even though the literacy rate has doubled over the past 20 years, the number of illiterates has actually increased by over 80 million during that period due to the high population growth rate. There is a serious shortage of qualified teachers, which grows worse as the population increases. In 1975, there were already about 500 primary school students for every qualified teacher.

It is clear that if India is to bring enough education to its rural population in order to mount an effective nationwide campaign to increase its food production and control the population growth rate in time to avert a national catastrophe, the traditional teacher-classroom concept of education must be bypassed, or at least supplemented, with modern techniques utilizing audio-visual aids. The most flexible and effective of those is modern television. But how can a developing country, without benefit of a nationwide network of ground TV stations, reach a large enough segment of a rural population with conventional education TV methods in time to meet its pressing needs and at a price it can afford? The answer clearly is the Direct Broadcast Television Satellite.

There is, of course, a very serious aspect of this marvelous new product of science that can bring education and enlightenment to the illiterate and poor of a vast country. Can we— or can the Indians themselves for that matter—really assess the impact of bringing, within the time span of a single generation, literacy to half a billion people? What changes will that cause to the social and political fabric of a predominantly rural country? Will there be enough challenging job opportunities for the next educated generation? If not, will there be widespread frustration? Will the Indian population as a whole, known for its calm acceptance of fate and the individual's station in life, become more restless, with resulting turmoil and bloodshed?

It seems we are right back to my earlier question: Should we put Prometheus in chains? Clearly, here again, science and technology help, carrying the fire of knowledge and enlightenment about the rest of the world to people who had not known it. I am personally deeply involved in this particular satellite and the India educational experiment. Many of my friends and associates, who for many years have worked in this program with unbounded enthusiasm because of its epoch-making humanitarian potential, have quietly confided me their concern and moral scruples, as no one can fathom the long-term consequences of such an educational revolution.

Speaking for myself, I hold the strong belief (and so told my friends) that now that science and technology have given us the capability of bringing education to the illiterates in the world, we have no moral right to deny it to them. Maybe the fact that I spent twenty years of my life in Huntsville,

Alabama has something to do with that conclusion. When I arrived there in 1950, the opinion was widespread among the predominantly white population of Huntsville, that most black children were either not ready for first-class education, or that a general raise in their educational standards would only lead to trouble with job placements. Today, Huntsville has some of the finest integrated schools in the nation and few, if any, of the predicted problems have arisen.

4.2 Environmental Survival

The 3½ billion astronauts not only deplete the dwindling resources of spaceship Earth like drunken sailors, they also poison their life-support system as though they were implementing a global suicide pact.

Many of our environmental problems can be solved with existing scientific knowledge and technology, provided we are willing to pay the price for it and ready to sternly enforce present and future water and air pollution and waste disposal laws. These laws frequently collide with the economics of a factory, or the realities of a community budget, and we must probably be a little patient before all our rivers are again safe to swim in, full of fish, and smog will be an unpleasant memory of the past. But as far as the United States is concerned, I am confident that we shall see drastic and obvious improvements within the next decade.

It is almost ironical that some of civilization's most applauded accomplishments have become the most serious threats to our environment. The earliest cities in antiquity were built on river banks because of the convenience the river offered as combination sewer and source of fresh water supply. The river as a convenient sewer led to the invention of the flush toilet, which some environmental experts in modern sewage treatment now consider one of the most disastrous inventions ever made by man. It led to the habit of committing not just our body wastes, but most of our other wastes as well, to sewage water, which we now find we must remove from that water in expensive sewage treatment plants again before committing it to the river. In retrospect, an earlier invention of the chemical toilet and of energy-producing incineration of dry garbage may have been a better overall approach.

During the early days of the industrial revolution, a belching factory chimney was a visible sign of progress and the pride of the community. In most developing countries it still is. Today, in the United States any plan for a new factory or public utility plant with smokestacks is subjected to rigorous reviews of the scrubbing equipment, and often never materializes because of local aesthetic or environmental objections.

Collisions between the requirements of those responsible for meeting our ever-increasing electric energy demands and those who are to protect our environment can be expected to continue for years to come. While a few other industrialized nations established effective mechanisms to resolve these issues before we did, the United States now has set up machinery that makes me confident that we shall see a steady rate of improvement of our environment.

Let us just take a quick look at how this machinery works. In 1970, by a Presidential Act, the Environmental Protection Agency was established to formulate Federal Standards spelling out in chapter and verse the type and amount of permissible pollutants that may be added to our air, rivers and lakes. As thousands of factories and service operations, ranging from steel plants to car wash facilities, and millions of automobiles, were outside the new legal limits, target dates were set by which everybody had to comply or be in violation of the law. This new Environment Protection Agency was given the authority to police the progress and take the violators to court.

Under the Act, cleaning up our air and water will actually be a step-by-step process. The rationale behind this approach was that applying new yardsticks to industries employing millions of people is a painful and time consuming process. While great improvements can often be made at moderate cost, true perfection is always costly. Thus a set of target dates, spread several years apart, was established under which the screw was tightened gradually and the permissible amount of pollutants reduced in steps.

Law enforcement, however, is only one side of the problem of environmental survival. It clearly also has a scientific aspect which touches more directly on the topic of my talk. In fact, the protection of our environment has so many scientific facets that I must limit my discussion to two examples and what remedial steps could be taken through responsible scientific investigation and application.

My first example involves water pollution. Our rivers, lakes and estuaries are populated by thousands of species, from microorganisms to oysters and game fish, which feed upon each other in an extremely complex food chain. This chain produces a snowballing effect with the result that some organisms accumulate many hundred thousand times the pollution concentration of the water in which they live. Stories you read about oyster beds being closed by the Food and Drug Administration, or mercury-contaminated fish being taken off the market are based on this multiplication phenomenon.

The sorry fact is that we understand the interaction between pollutants and marine life only in a very sketchy fashion. More scientific research in this field, supported by a well-equipped marine laboratories and research ships, should be accorded highest priority. As Captain Jacques-Ives Cousteau has pointed out, river pollution is even endangering the survival of the Earth's continental shelves, which support most of the life in the open oceans.

Now as to remedies; permit me to become a little parochial again and tell you how the space program can help. The two Landsat satellites I talked about earlier have detected numerous hitherto unknown sources of river and lake pollution. Their images have been successfully used as evidence in court action against offenders. They have also detected oil slicks in mid-ocean and identified the tankers whose skippers decided to clean their bilge under cover of night.

Satellites as monitors and protectors of our environment lead me to my second example, which involves air pollution. You have undoubtedly read about the controversy raging on the subject of the atmosphere's protective ozone layer, and whether or not supersonic airliners or the Freon gas escaping from spray bottles may decompose that layer. Statistics, little contested, show that any reduction of the layer's ozone content increases the amount of solar ultraviolet radiation which reaches the Earth's surface, and with it the statistical occurrence of skin cancer. One of our satellites now has been monitoring the Earth's ozone layer on a global scale for several years. It has identified numerous occasions where a sudden burst of protons emanating from the Sun did indeed produce a noticeable drop in the layer's ozone content, but the old balance, itself produced by solar radiation, was soon restored. Only a single manmade phenomenon was detected that produced a similar if somewhat lesser effect: It was a high-altitude explosion of a French nuclear bomb over the Pacific, a test program meanwhile terminated by the French government. Evidence of any effects in the ozone layer caused by supersonic aircraft, which are in widespread use by air forces all over the world, or by Freon spray bottles, has not yet been found. But let me hurry to add here that I am not ready to conclude that there may not be some problem after all. Research in this area should continue under high priority.

Another field which, in my opinion, deserves much more attention, priority and resources is that of marine biology. You can study the food and behavioral pattern of a large mammal pretty accurately in captivity or in a wildlife preserve. But when it comes to the question how the thousands of pollutants and toxic metals that come down with the river water are absorbed by the thousands of different marine organisms, ranging from microorganisms to oysters, shrimp and game fish, we have to establish more marine laboratories in more estuaries to study these processes in situ. Being affected by tides, temperatures, silt, soil, speed of water movement and what have you, the conditions differ from place to place and simulation in test tanks and aquariums has turned out to be hopeless.

Before I leave the subject of environmental survival let me leave you with some food for thought, and maybe a smile.

Our rightful concern about the environment has led to the legal requirement in the U.S. to file an environmental impact statement for practically every major new undertaking. This requirement may be easy to applaud, but it is not always so easy to implement.

Suppose the elder Henry Ford, before committing himself to mass produce his Model T, would have had to file an environmental impact statement. In his contagious optimism, that statement would undoubtedly have included the highly desirable effect of personal mobility for every citizen and the ensuing subsequent public demand for better roads. A critical reviewer of Mr. Ford's first draft may have added the unpleasant odor of exhaust gas, but it is not likely that that particular aspect would have bothered the authorities. Had that reviewer also added that the automobile would become so popular that in 1975, in America alone, over 46,000 people would get killed and over 2,000,000 people injured in car accidents, no government appraiser of that environmental impact statement would have believed that figure, anyhow.

But I am sure that both Mr. Ford and his critical reviewer would have missed what was probably the most important item in their impact statement: the fact that the automobile would rid America cities of the housefly. As a carrier of contagious diseases the housefly is one of the worst killers known to man. Its home and breeding ground in the horse-and-buggy days was the horse stable in the city, and it fed on what it found in the horse manure on the streets. No one can deny that the automobile drove the horse, and with it the horse stable, out of the American city. It thus surely made the cities healthier. True, we traded this gain for the health-impairing effects of smog, but that will probably be a temporary nuisance which with cleaner cars will soon disappear.

The moral of my little story on the effect of the automobile on the housefly is that our crystal ball may often not be clear enough to predict the long-range environmental impact of an innovation.

4.3 SPIRITUAL SURVIVAL

The often colliding demands between those expected to supply mankind with its insatiable needs of material resources and energy and those responsible for the protection of the Earth's life-supporting biosphere are bound to lead to an ever-increasing regimentation of many aspects of our lives. Not one of the many candidates for the 1976 presidential election failed to mention the undesirability of the continuous growth of the Washington bureaucracy in his campaign speeches, and every one was applauded when he pledged to stop or reverse that trend. After all, one of the basic premises of the founding fathers of our great Republic had been that the best government for the newborn nation would be the smallest government, which would interfere as little as possible with individual freedom and initiative. Yet, as we are celebrating the Bicentennial, we find the number of people on the federal payroll far exceeding the total population of the United States at the time of its birth.

Having served in the U.S. Civil Service myself for over 20 years, I always resented the widespread belief that government employees are some sort of drones in near-permanent hibernation, which show some temporary signs of life only when presented with their bimonthly paychecks. I have also never shared the view that government has grown so big mainly because it is so difficult to lay off civil service employees, or because they cling more tenaciously to their jobs than other people. I have met the finest, most capable and hard-working people in the U.S. Civil Service, and can assure you that Americans would not have landed on the Moon without their leadership qualities and dedication to what most of them considered a great patriotic challenge.

Why then is it that "Big Government" is in the doghouse? Let us look at the reasons that made Government grow as big as it is. Two hundred years ago, the U.S. consisted of 13 states. Today it has 50, while the total population grew from a little under 4 million to nearly 216 million. But equally important is that life in our highly-industrialized civilization has become infinitely more complex. A whole string of unprecedented regulatory or public safety needs arose which simply could not be turned over to the private sector. George Washington had no need for government agencies such as the Federal Communication Commission that would allot radio frequencies or issue permits and operating rules to the telephone, television and radio industry. Nor did he need a Federal Aviation Agency to operate airport towers and a nationwide air traffic control system with radar-equipped centers and thousands of navigational aids.

Unfortunately, government agencies, like human beings, have a tendency to put on a little fat as they grow older, and to slow down their pace. By the same token, young agencies, charged with exciting and difficult new tasks, display more vigor and vitality. The youngest members of our ever-growing fraternity of U.S. government agencies are the Energy Resources Council, the Energy Research and Development Agency and the Environmental Protection Agency. Obviously they play a vital role in the protection of two of the five pillars of our priority dome-resources survival and environmental survival—and should therefore be welcomed with applause.

But the American public has just become weary of any further growth in our federal bureaucracy, no matter how noble their objective or how pressing their need may be. For one, John Q. Public, grouches about the higher taxes he feels he will undoubtedly have to pay in one way or another to support any new agency. Even more importantly, he shudders at the relentlessly creeping growth of government interference with his business affairs and his private life. That, undoubtedly, is the underlying cause for the widespread interest in the Congressional hearings on the CIA and FBI—two types of institutions without which no large country can service in this dangerous world. It is also the main reason why any pledge of a presidential candidate to lower the boom on Big Government always assures him a big hand from the audience. Americans just have a profound dislike for anything that smacks of big Brother or Daddy Knows Best. Instinctively, they have a deep-seated concern for our spiritual survival.

A mere hundred years ago, this was a nation of open frontiers and unlimited possibilities. Many people have told me with a tone of nostalgia in their voice, that with the end of the western pioneering days the wind has gone out of the sails of the United States. They say a man can no longer build up his own life as he pleases—that he is being trained, tutored and manipulated from cradle to grave. Now, as a native European, I must admit that I do not see it all that glumly. There are still vast untapped opportunities, even for rustic pioneers and settlers, in many thinly populated parts of this huge country. But I concur that even in the vastness of the American west and southwest the prospective pioneer and settler of 1976 is bound to collide with the plans of some energy or mineral-related agency that wants to develop and exploit a local natural resource, or some environmental protective group desirous to stop local settlement and economic development altogether because it interferes with their plan for a wilderness area or a National Park. Nevertheless, in this respect American is still a lot better off than much of the rest of the world. In countries with a high population density there are usually no undeveloped areas left at all and opportunities for a rustic pioneer to start his own homestead from scratch have been non-existent for centuries.

The threat of an imminent collision between an expanding world population and the limited space and resources of a finite planet has given rise to the battle cries of "Limits to Growth" or even "Zero Growth." As [with] so many new ideological concepts that feed on a grain of truth, "Zero Growth" has been enthusiastically embraced by many. And yet, in its simplistic form, I am convinced it should be rejected as half-cocked and downright dangerous.

There is no question, of course, that some growth trends should indeed be of great concern, but continued growth in other areas is a vital necessity for the survival of humanity. Better land use, prudent family planning, protection of our environment, and conservation of energy and all of the Earth's precious resources are vital necessities that we should whole-heartedly endorse.

On the other hand, growth in food production, development of our hardly-tapped ocean resources, reduction of human suffering by provision of better health care in remote areas, or improvement of the worldwide level of literacy are areas in which we should support a continuous and vigorous growth. To subject ourselves to drastic across-the-board curtailment in production of necessary goods and services, I think, is unacceptable, both from the practical and spiritual viewpoint. The indiscriminate philosophy of "No Growth" makes no difference between growth of population and growth of means for continuing survival. We must reject this simplistic concept. Indiscriminate "No Growth" would remove the good with the bad, and in the process condemn that part of the world now living below subsistence levels to eternal poverty and misery, a brutal policy that would deprive man of hope.

In connection with my remarks on Resources Survival I mentioned the capability of our new Landsat satellites to monitor the worldwide status of food crops and to find new mineral and fossil fuel sources. These satellites can also be used as survey tools to make better use of the many unused parts of Earth. Central Brazil, for instance, is almost uninhabited, while Brazil's coastal areas are densely populated. Java is bursting with people while adjacent Sumatra and Borneo's population density is exceedingly low. It is neither poor climate, nor governmental desire that prevents the migration of people to these empty areas. It is lack of communications, lack of educational facilities, and lack of detailed knowledge of local soil qualities and mineral resources that slow down these highly desirable transmigrations. Multi-spectral surveys from orbit and communications satellites can effectively remove the obstacles.

But the spiritual survival of mankind has also an aspect completely unrelated to Big Government or the limitations set by the Earth's resources. It is more fundamental. In discussing the motivation of scientific research I said that the mainspring of science is curiosity. But curiosity has also been the driving force behind man's great exploratory voyages. We may well doubt whether Columbus, on the quarterdeck of the Santa Maria, was animated by a burning desire to reduce the freight rates on Indian tea when he sought a new route to Cathay and found America athwart his hawse.

What drives man to explore the unknown? Why are we flying to the Moon? What is our purpose? What is the essential justification of the exploration of space? The answer, I am convinced, lies rooted not in whimsy but in the nature of man. Let me give you my personal view.

Man, as a biological species, is a rather anomalous animal. Whereas all other animals establish a place for themselves in nature's highly cooperative and competitive ecological system, man has established his place in nature by altering his natural environment through such actions as practicing agriculture rather than eating the natural fruits of the trees and the plants, and clearing forests to build cities.

Whereas all other living beings seem to find their places in the natural order and fulfill their role in life with a kind of calm acceptance, man clearly exhibits confusion. Why the anxiety? Why the storm and stress? Man really seems to be the only living thing uncertain of his role in the universe and in his uncertainty, he has been calling since time immemorial upon the stars and the heavens for salvation and for answers to his eternal questions: Who am I? Why am I here?

Wherever he fought, he invoked the stars for help. Wherever he loved, he invoked the Moon. And all great religions hold out eternal life and salvation as man's reward for his good deeds here on Earth.

Whereas most animals follow, for their survival, certain telltale scents which are too refined for human perception, man seems to be uniquely equipped to perceive certain vibrations emanating from the celestial environment. As a result, astronomy is the oldest science, existed for thousands of years as the only science, and is to this day considered the uncrowned queen of the sciences. Although man lacks the eye of the night owl, the scent of the fox, or the hearing of the deer, he has an uncanny ability to learn about abstruse things, like the motions of the planets, the cradle-to-the-grave cycle of the stars, and the distance between the galaxies.

Whereas all other species seem resigned to the environments in which they have been born, man clearly [is] not. Since his early beginning he has wanted to fly, and today he does fly.

What is man's motivation? Why does he always want to explore what is outside his abode? Why is he so eager to pioneer activities beyond his natural endowments?

I guess it is all just in the basic makeup of man as God wanted him to be. And it explains why, now that all the white spots have been removed from our maps and nothing seems to be left to discover on the surface of the Earth, for the first time in his long history man has developed the uneasy feeling that there are no more frontiers—forever.

Behind all the talk about "Zero Growth" is really the fear that the species of man is heading for the fate of a domesticated animal confined to the fences of its pasture. Moreover, as our numbers keep increasing, and less and less elbow room will be left for the individual, there looms the dreadful aspect that we may ultimately be reduced to mere numbers in a coercive and tightly regimented society like that of the honey bee. There would be no spiritual survival and personal freedom would be dead.

This horrible outlook of the doomsday prophets for our long-range future leads me smoothly back to my favored subject, space flight. I happen to be convinced that man's newly acquired capability to travel through outer space provides us with a way out of our evolutionary dead alley. It takes, as it were, the lid off the pressure cooker called Earth. Who can honestly claim that the Earth is the only place where man can live? Our astronauts ran around happily on the Moon, and whenever the distances they had to travel on the lunar surfaces became too great for walking, they even used a rental car they had brought along. When they became uncomfortably hot or cold, they just made a slight adjustment in the thermostat setting of their pressure suits.

If someone asked Mr. Hilton to build a hotel on the Moon, offering the same comforts of his Earthly inns, we could give him the technology to do it. We could even feed his guests with food grown in pressurized green houses and chicken farms adjacent to his hotel. Creating permanent habitats on the Moon is no longer a question of science or technology, it is merely a question of our will and determination to do it.

As long as the flame of man's innate urge to extend his arena of activities and his knowledge is not extinguished, there is no reason why we could not extend our domain beyond the Earth to the entire solar system. Of course, we do not have the technology to do all of it at once, but we can start slowly and implement the program gradually during the remainder of the five billion years the planet Earth is expected to remain climatically inhabitable.

We hear the argument that on the Moon there is no breathable atmosphere. Well, there is none in Los Angeles either, yet people still manage to live there, and many live there very well indeed. Mars' atmosphere, we are told, is too thin for human habitation. So what? People could live and work on Mars in pressurized, climate-controlled buildings like they do in Alaska or Siberia today. Venus is too hot! Well, suggestions have been made to inject loads of a suitable breed of algae, carried there by a fleet of rockets, into the higher, cooler, and water vapor-containing layers of the Venus atmosphere. The algae, the originators of this scheme claim, would multiply by leaps and bounds and through photosynthesis soon convert Venus carbon-dioxide atmosphere into one containing oxygen. The removal of the carbon-dioxide would destroy the greenhouse effect, with the result that the surface temperature of Venus would cool down to a level that would make the planet habitable for man.

Now, I do not think that at this moment anyone can vouch for the practicability of this concept, but then, who can predict what man will be able to accomplish in the next five billion years? With unlimited solar and thermonuclear energy at our disposal, we may learn to re-engineer even some of the many Moons of Jupiter and Saturn and make them fit for human habitation.

For the more immediate future, the 21st century, another concept for an unlimited growth of humanity has recently drawn much serious attention. Professor Gerard K. O'Neill of Princeton University has suggested the building of huge manmade habitats orbiting the Earth at the so-called Libration or Lagrange points, which are equidistant from the Earth and Moon. A typical habitat may be a hollow pressurized cylinder 19 miles long and 4 miles in diameter, and would accommodate 10,000 people. It rotates slowly about its axis to generate Earth-normal gravity through centrifugal acceleration. The cylinder mantle consists of alternate bands of glass and materials supporting interior land areas. The land areas, totaling about 100 square miles, provide housing for the inhabitants in surroundings more desirable than at most inhabited places on Earth. They are covered with soil supporting the growth of trees, lawns and gardens. From those land areas a resident would see a reflected image of the disc of the Sun in the sky, and the Sun's image would move from dawn to dusk within the time span of a day, just like on Earth.

Agriculture and poultry farming to support the space colony would be carried out in one or several separate cylinders nearby with climate control for optimum growth. Industrial activities, which provide employment for the colony, would likewise be conducted in several independent modules in order to avoid any kind of industrial pollution problems. Much of the industrial processing would be done taking full advantage of zero gravity. The main industrial products of a colony would be other habitats for future generations of space colonists. Rough estimates indicate that a population of 10,000 could manufacture about one new colony of equal size every two years. In addition, it could build two orbital solar power stations of 5,000 megawatts each to beam badly needed energy to the Earth. This could easily become the most attractive way to raise more money on Earth for additional expansion in space.

Thus, a mechanism can be generated through which the goal of further unmanageable population growth on the limited Earth is actually reconciled with the concept of "Humanity Unlimited."

Professor O'Neill suggests to build the first habitats from lunar material launched to the Lagrange points with an electromagnetic launching device located on the Moon. Powered by a nuclear reactor, this "mass driver"—in essence a recirculating linear electric motor—can accelerate a big bucketful of unprocessed lunar surface material every 150 seconds to the relatively low lunar escape speed required to reach the Lagrange point. A typical Apollo sample of lunar soil contains about 40 percent oxygen, 30 percent metals and about 20 percent silicon, which is just about the right mix for the metal, glass, soil and atmosphere required to establish a habitat. Calculations show that it is more cost-effective to process the lunar raw material in the zero-gravity environment at the habitat site itself, rather than on the lunar surface.

Once, many millennia from now, we will have run out of usable lunar material, Professor O'Neill proposes to bring the material in from the asteroid belt. The total quantity of material within only three of the largest asteroids is quite enough to permit building space colonies with a total land area more than ten thousand times that of the Earth! With the inexhaustible sources of raw material from the asteroid belt and of energy from the Sun, it seems that all that has to be added is human will and brain power to assure humanity virtually unlimited growth.

I stated a little earlier in my speech that I believe that the mere fact that something is scientifically possible does not necessarily make it permissible. Are we possibly stepping beyond the bounds of human destiny by setting out to populate the vast empty spaces outside of the Earth?

I do not think we do, and I offer you three reasons. For one, nobody gets hurt in the process, but hundreds of millions would either starve to death, never be born, or live under unbearable regimentation if we decided not to pursue the option of a mass exodus into space.

Secondly, I think the concept of multiple space colonies can relieve us of a lot of problems here on Earth. Since the beginning of recorded history man has conducted all sorts of social experiments. Time and again we see the spectacle of a charismatic leader arousing millions for some political, racial, economic, religious or nationalistic cause which subsequently is pursued and defended in bloody wars. In the age of space habitats this permanent danger source could be removed from the Earth. Anyone with a sufficiently large following could henceforth open up his own habitat, where his group may lead the kind of life it prefers, without outside interference and the urge or means to impress its own lifestyle on others.

Do not think that this is in the realm of utopia. Right here in the United States a development took place along quite a similar pattern. Large affinity groups of people who had been persecuted in Europe for any number of reasons came to America with the sole motivation and hope to be able to continue the kind of life they preferred: the Puritans, the Amish, the Mormons, the Hugenots, the German Democrats of 1848, to name just a few. The vastness of America, provided adequate isolation between these groups, which prevented them from promptly getting into each others hair. In outer space, the separation between the habitats will have pretty much the same effect. Travel between the various habitats, however, should be easy, speedy, and cheap, as all of them will circle the Earth at the same orbital altitude and speed, thus very little energy must be expended to go from one to another. Cultural and social exchange between habitats should therefore be easier than between different nations today. Any attempt of the population of a single habitat to build up the capability to act against one of its neighbors could be easily nipped in the bud by a roving international police force.

Thirdly, I strongly believe in what Immanuel Kant would have called the *Cosmic Categoric Imperative*. I think a strong argument can be made that it is indeed the Creator's intent for man to use his intelligence to extend his domain beyond his home planet. Let me elaborate a bit.

Bioscientists have learned that for the most primitive forms of life to come into being anywhere, rather narrow constraints of physical conditions must prevail. For instance, there must be some water, temperatures must be within certain limits, and certain toxic substances must be absent. After, under a conducive set of conditions, simple life finally got entrenched, several billion years of a reasonably stable environment was required for the further evolution of intelligent life.

Astronomical and exobiological research during the past 25 years, aided by a fleet of unmanned planetary spacecraft, has led many to the conclusion that the necessary suitable environmental conditions prevailing over a sufficient time span to spawn intelligent life probably exist only on a relatively small percentage of the planets in our galaxy. But as we behold the immense machinery of the cosmos, with its inexhaustible sources of energy and raw material, one thought is irrepressible: could it be one of the fundamental concepts of the entire creation, that wherever, in spite of these odds, the evolution of higher intelligence has been successful, the Creator expects that form of higher intelligence to expand its activities from its home planet to its cosmic neighborhood?

As the great Russian space flight pioneer Edward Ziolkowsky put it, three quarters of a century ago: "The Earth is the cradle of man. But who wants to stay in the cradle forever?"

4.4 Nuclear Survival

In the field of nuclear strategic capability, the two main adversaries, the U.S. and the U.S.S.R., now clearly possess a capability of multiple overkill. The destructive power has become so immense that in any all-out nuclear exchange both sides would perish. The situation has been called a Balance of Terror. In spite of one or two uneasy confrontations and several non-nuclear wars (in which both sides were deeply or at least peripherally involved) that balance so far has proven to be remarkably stable.

One is reminded of that famous animal behavior experiment which showed that two male scorpions locked in a bottle will not harm each other, although in the open they are known to engage in combat ending invariably in the death of one.

In the confinement of the bottle, neither scorpion will attack, as it seems to be aware of the fact that, even if it struck first, there would be enough fight left in the dying opponent to retaliate against its trapped attacker with a deadly sting.

However, soon after the public had begun to draw a degree of comfort from the idea that nuclear double suicide was simply too terrible to even be thinkable, there were signs of possible destabilizing effects in the uneasy Balance of Terror. Suppose one side "hardened" (missile jargon for strengthening against near-misses of atomic bombs) its missile sites while the other did not? Or suppose only one party acquired an antimissile system that could reliably protect its missile sites, as well as its cities and industrial complexes?

When the Cold War showed signs of thawing in the late sixties, the United States and the Soviet Union decided to talk things over. Both sides readily agreed that an unchecked nuclear missile race was fraught with enormous dangers, and painfully drained the economies of both countries. But from an agreement on a basic truth, that was so obvious that it almost became a platitude, to an ironclad and cheat-proof arms control agreement was—and is—a difficult and rocky road. Delicate questions such as mutual inspections of top secret missile sites arose. What good was an agreement on missile numbers if both sides were free to increase the "band" delivered by the warheads or to place several warheads into the nose of a single rocket? Suppose that these multiple warheads could even be independently targeted and were maneuverable to evade antimissile missiles? If one of the two sided decided not to keep up this endless race, would it become a "sick scorpion," possibly tempting its bottle mate to use the opportunity to strike, for fear that it may one day become sick itself and an opportune victim of the other?

This is the eerie backdrop of the SALT (Strategic Arms Limitations Treaty) meetings, possibly the most fateful peace-seeking gatherings in which man has ever engaged. For the moment, it looks as if the huge rockets with their awesome hydrogen warheads are keeping the big powers at peace in spite of their many disagreements and conflicting interests. Only the future can tell whether, as we all hope and pray, the age of the great world wars will be a thing of the past—forever.

We have a dramatic example here, how intricately science, technology, military policies and national resolve to survive are interwoven. Clearly, life-and-death questions involving the fate of our nation—any nation—can not be answered by scientists and technologists alone; they will always be taken up at the highest policy-making level of a country.

However, it was the international fraternity of nuclear Prometheus's—some of the most brilliant scientists of our age—who had handed their respective national leaders the hot potato of that new unprecedented atomic fire with all its attendant hazards. The nuclear scientists, and the rocket scientist's who furnished them with the means of transporting their atomic fire to distant targets, must leave no stone unturned to now also help their national leaders harness that fire.

The value of any existing and future SALT agreements is directly related to the ability of each side to monitor the other side's adherence to the agreements. This is an area where rocket and space technology at this very time is making a vital contribution to mankind's military survival. Without the capability of U.S. and Soviet reconnaissance satellites to keep an alert eye on activities on both sides of the Iron Curtain—such as new missile launching sites, construction of missile-carrying submarines, or massive troop movements—any past and future SALT agreements would lack credibility with the two parties to the treaties and not be worth the paper they are written on. Both sides see eye to eye that agreements involving their military survival must be cheat-proof. It is significant that the SALT 1 treaty contains an explicit provision that neither side shall tamper or interfere with the other side's inspection satellites. This is a clear expression from both sides that all scientific and technological efforts aimed at improving the mutual inspection capability will greatly contribute to a safer world.

Nuclear survival also depends on our ability to learn to control the proliferation of nuclear fuel.

All industrialized nations (and those who want to become industrialized) have an insatiable appetite for energy. If they have no reliable access to coal or oil, they want nuclear electrical power. But every nuclear reactor, whether of the conventional "light water" type, or the futuristic "fast breeder," creates as a byproduct plutonium, which can be extracted and fashioned into atomic bombs.

According to projections by the International Atomic Energy Agency in Vienna, Austria, the worldwide total of nuclear power reactors will quadruple to 800 over the next 100 years. Studies estimate that by 1990 nuclear reactors of less-developed countries will generate 30,000 pounds of plutonium yearly. That is enough to make 3,000 small atomic bombs.

In a recent Senate hearing it was revealed that by mid-1974 the United States, which supplies about 70 percent of the world market in this field, had exported $3.9 billion in nuclear materials and equipment. Current exports were estimated at $1 billion a year and rising sharply.

The United States sells nuclear fuel "rods" to foreign countries only with the stipulation that the "irradiated rods" must be returned to the U.S. As they are highly radioactive after several years of neutron bombardment in the reactor, it is impossible for a country without very sophisticated equipment to extract the plutonium which formed in the rods during the activation stage of the reactor. But we are not the only supplier in this field, and others have more lenient rules for their sales.

It is hardly surprising that leading nuclear physicists are alarmed and recommended in congressional testimony a complete export ban on certain critical equipment such as plutonium extraction systems, isotope separation plants, or technology helpful in the perfection of the still highly experimental breeder reactors. The truly worrisome question is, of course, whether the horse isn't out of the barn already. The thought of nuclear bombs in the hands of terrorist groups or madmen is frightening indeed.

I believe the space program can offer a powerful tool to monitor and control the proliferation of nuclear fuel. A recent study proved the feasibility of establishing, for a fraction of the cost of our yearly nuclear exports, a satellite system that could identify and locate every nuclear fuel rod anywhere on Earth.

Let me inject here a few technical details, so I can better explain the concept. A fuel rod is a long tube filled with pellets of metallic nuclear fuel. For a reactor to produce heat, it is loaded with a sizable number of these rods, all separated from one another at carefully calculated distances. In a light-water reactor the space between the rods is filled with water, which heats up when the reactor "goes critical." The heat extracted from this water then provides the power for the turbo-generators that produce electricity.

After several years of exposure to the intense neutron bombardment inside the reactor, the fuel rod contains such a high percentage of fission products that the nuclear chain reaction gets choked in its own nuclear waste. The highly radioactive rod must then be removed from the reactor and replaced by a new one. The "irradiated rod" can either be placed in a long-life coffin and buried (usually under water), or it can be sent, coffin and all, to a remote-controlled chemical facility, where the fission products are extracted. It is in this extraction process that the plutonium accrues.

Coming back now to the space monitoring concept, the idea is to attach to each fuel rod, while it is outside of the reactor, a wristwatch-sized receiver/transmitter, or a "transponder", which, when interrogated by a powerful satellite transmitter, will return a signal revealing its precise location and serial number. If the device is removed from the rod or deactivated by anyone, the satellite's bookkeeping computer would at once record the tampering or the disappearance of the signal.

Such a satellite monitoring system can be built with existing technology. Of course, it would provide effective protection against uncontrolled proliferation of nuclear fuel only if the world's few manufacturers of nuclear fuel can be persuaded to participate in the program by labeling all the rods they sell. Knowledgeable people in this field have told me that the concern about worldwide uncontrolled proliferation of nuclear fuel is so universal, and just as serious to the Soviets as it is to us, that they thought it would not be difficult to conclude an international treaty on this subject patterned on the ban on atmospheric testing of nuclear bombs.

Any attempt to maintain peace with a strong adversary who is in an expansionist mood can be successful only if the negotiations are conducted from a basis of strength. I am still convinced that World War II could have been avoided if, in 1939, Great Britain had sent a man of Churchill's bent, escorted by some British and Allied military leaders, to Munich, instead of a man with an umbrella willing to settle for what he called "peace in our time."

American SALT negotiators should be armed with better weapons than umbrellas. Their opposite numbers should know, for instance, that we are building up an effective globe-girdling system to monitor the movement of Soviet missile-carrying submarines, that we have a new breed of fighter aircraft that can quickly gain and maintain air superiority in any area of confrontation, or that we are making effective preparations to be able to stop a massive non-nuclear armored onslaught of the Warsaw Pact powers against the NATO forces without being forced into immediate nuclear retaliation.

The contributions required from science and technology to protect our pillar of nuclear survival are manifold indeed. In the wake of the unfortunate Vietnam War scientists and engineers working in the field of advanced weaponry were subjected to much harsh criticism. They found their moral standards questioned and were actively encouraged to discontinue work in these fields. In the last analysis, of course, each individual must settle such questions with his own conscience and I would not label a man a traitor who feels he must refuse to work on anything involving weapons of war. Speaking for myself, I hold to the conservative belief that until that day when all people have become saints and all swords have been beaten into plowshares, the risk of a nuclear Armageddon will be greatly reduced as long as Americans keep their Armed Forces the most advanced in the world.

4.5 SCIENTIFIC SURVIVAL

There can be no doubt that the present mood of many Americans is anti-scientific and anti-technological. Science and technology, they seem to feel, have brought us more problems than blessings. By the same token, they realize that a return to the simple life, the "good old days," is not possible, either. The family-owned farm, for instance, nostalgically remembered by many a city dweller as the place where he was born and raised, may have been absorbed by a giant "agro-business," where on an immense spread huge crops are harvested with a handful of people operating expensive machinery.

Thus, with the retreat to their rustic past blocked, and life in the cities getting less and less pleasant, many Americans view the future with dismay. But while they continue to blame science and technology for many of the ills that have befallen our society, they grudgingly concede that probably only more science and technology can get them out of the mess they feel they are in.

I am certain that the tide of anti-science sentiments is about to turn, and that, as it surges back, ever-stronger demands will be heard for constructive contributions of science and technology to the world's down-to-Earth problems. The near-simultaneous creation of the Energy Research and Development Agency and the Environmental Protection Agency, both at a time when everybody is complaining about the never-ending growth of Big Government, is a pretty convincing indication for the reversal of the tide.

Like the magician in the vaudeville show, the scientist is once again expected to pull the white rabbit out of the hat and cure our ills by working another of his miracles. But before you can pull a white rabbit out of a hat, someone first has to place the rabbit into the hat. This is the function of basic science.

In assessing the merits of any proposed experiment in basic, as contrasted to applied, science, one must carefully avoid the criterion of immediate payoff. You probably know the famous story of the British Prime Minister visiting Michael Faraday in his laboratory. As Faraday explained his experiment on electromagnetic induction, the Prime Minister demanded what practical benefits one could ever expect from this eerie sort of thing. Faraday replied: "I do not know, sir, but I am sure that you will find a way to tax it." A century later, a worldwide electrical industry had grown on the foundation laid by Faraday's work.

Looking around at the many facets of scientific pursuit in 1976, I can discern many opportunities just as promising as Faraday's work. We touched already on the field of marine biology and its crucial importance for our environmental survival. We mentioned nuclear fusion research as the great hope to satisfy mankind's insatiable appetite for energy without the risks of nuclear fuel proliferation. We have seen how, in recent years, solid state physics has revolutionized the field of electronics and global communications, and realize that we have only taken a glimpse at the fantastic future world of microminiaturization. We watched the fabulous successes of medical researchers who virtually eradicated most of the contagious diseases and mass epidemics of the past, while we are still witnessing their frontal assault against a few remaining and stubbornly resisting killers, such as cancer and heart failure. And we have seen how, through the "green revolution," the breeding of hardier and superior species of crops, the desperate need of the growing population for food, has been alleviated in many parts of the world.

These examples should remind us that science is indeed helping mankind in many ways. But there is no denying that the same scientific accomplishment that helps mankind in such a large measure often also causes great harm. Take DDT, the pesticide widely used in agriculture the world over. Ecologists tell us that as the rains wash DDT from the fields it is supposed to protect into brooks and rivers, it creates havoc with the entire complicated food chain of marine organisms. Moreover, it decimates many bird species feeding on the crop-endangering insects dead or dying from DDT.

Shall we then, as many voices demand, simply ban DDT? It has been estimated that if India would impose such a ban, the crop losses due to insect infestation would lead to the starvation of many millions of people every year. A better solution, and from what I hear one we can realistically expect in the very near future, is for science to come up with a DDT replacement that decomposes after a few months when dry and at once when diluted with water.

Even people with an anti-scientific bend will probably agree with this kind of scientific effort, but they still get up in arms when they hear that some of their tax money is being spent on science objectives that they are unable to relate at all to the benefit of man. As an example, let us take planetary research with unmanned spacecraft. People know that this program costs many millions and wonder whether there are other scientific objectives that should be accorded a much higher priority.

Unmanned spacecraft have now visited Mercury, Venus, Mars, and Jupiter, and one of it is on its way to Saturn, the planet with the mysterious ring. The scientists found the four planets these spacecraft have already surveyed not only very different from the Earth, but vary different from one another. The surface pressure of Venus' carbon dioxide atmosphere, for instance, was found to be about a hundred times as high as that of the Earth. Mars' atmosphere, by contrast, turned out to be so tenuous that its surface pressure is only about one percent of ours. Giant Jupiter's atmosphere was found to be about 1,000 kilometers thick and covering a storm-whipped ocean of liquid hydrogen.

The atmospheres of each of these three planets also displayed extremely interesting and quite unique meteorological phenomena. Their systematic observation will tell meteorologists a great deal about some of the fundamental laws that make different types of atmospheres tick, our own as well as exotically different ones. This broadening of their understanding of meteorological processes in general is bound to improve their long-term weather forecasting ability. This, in turn, is of immense value to sea and air navigation, the tourist industry, and most importantly, to agriculture. In India alone, a substantial portion of the yearly rice harvest is lost every year due to the inability of meteorologists to precisely predict the arrival day of the monsoon rains for a given region of the vast subcontinent.

Science may be motivated by man's innate curiosity, and it most certainly provides man with the most reliable set of answers to Pontius Pilate's skeptical question, "what is Truth?" But in our fast-moving world, scientific survival is also a vital fifth support pillar of the dome representing the all-encompassing objective of "Mankind's survival as a Species."

Science and Religion

This Symposium has been sponsored by the Lutheran Church of America and my presentation would be incomplete if it failed to address the potential of the powerful forces of religion to help protect us from ourselves. I said earlier in my talk that neither science nor technology has a moral dimension. Can religions or their human institutions, the churches, give us the moral and ethical guidance we so desperately need to protect us from the genie that science allowed to escape from the bottle?

In the Middle Ages, Western man lived under stern ethical directives from the all powerful Catholic Church and there was little room or reason left for doubt. Its old ramparts of faith provided the believers with a feeling of protection, comfort and oneness with God. But ever since the Reformation and the Renaissance the Christian churches, already divided among themselves, have been battered by a relentless onslaught of scientific skepticism. Faith in the Holy Scripts gave way to unfettered curiosity, and doubt was raised about things hitherto accepted as unquestioned truths. This has led many of our contemporaries to believe that science and religion are not compatible, that "knowing" and "believing" cannot live side by side.

Nothing could be further from the truth. Science and religion are not antagonists. On the contrary, they are sisters. While science tries to learn more about the creation, religion tries to better understand the Creator. While, through science, man tries to harness the forces of nature around him, through religion he tries to harness the forces of nature within him.

Some people say that science has been unable to prove the existence of God. They admit that many of the miracles in the world around us are hard to understand, and they do not deny that the universe, as modern science sees it, is indeed a far more wondrous thing than the creation medieval man could perceive. But they still maintain that, since science has provided us with so many answers, the day will soon arrive when we will be able to understand even the creation of the fundamental laws of nature without a Divine Intent. They challenge science to prove the existence of God. But, must we really light a candle to see the Sun?

Many men who are intelligent and of good faith say they cannot visualize God. Well, can a physicist visualize an electron? The electron is materially inconceivable and yet it is so perfectly known through its effects that we use it to illuminate our cities, guide our airliners through the night skies and take the most accurate measurements. What strange rationale makes some physicists accept the inconceivable electron as real while refusing to accept the reality of God on the grounds that they cannot conceive Him? Being a physicist myself, I strongly suspect that, although we really do not understand the electron either, we are ready to accept it because we managed to produce a rather clumsy mechanical model of the atom consisting of electrons and other equally inconceivable subatomic particles, but we just wouldn't know how to begin building a model of God.

For me the idea of a creation is not conceivable without invoking the necessity for God. One cannot be exposed to the law and order of the universe without concluding that there must be a Divine Intent behind it all.

Many modern evolutionists believe that the creation is the arbitrary result of random arrangements and rearrangements of uncounted atoms and molecules over billions of years. But when you consider the development of the human brain within a time span of less than a million years, statistical studies cast grave doubt on the question whether this relative short time span just was really long enough for random processes to produce the brain whose tremendous complexity we are only now beginning to understand.

Or take the evolution of the eye in the animal world. What random process could possibly explain the simultaneous evolution of the eye's optical system, the nervous conductors of the optical signals from the eye to the brain, and the optical nerve center in the brain itself, where the incoming light impulses are converted to an image the conscious mind can comprehend?

There is the even more mysterious interaction between animals and plants. We know that the eye of the honeybee cannot see red, but is sensitive to a band of ultraviolet light which the human eye cannot discern. Flowers, depending on visits by bees for their pollination, have developed intricate ultraviolet patterns designed to attract the bee's attention. Did the flowers fashion the bee's eye, or did the bees—through limiting their visits to a few flowers which, through random mutation, happened to have these attractive ultraviolet patterns—ensure their survival as the fittest?

Let us be honest and let us be humble: can all this really be explained without the notion of a Divine Intent, without a Creator?

Also, it is one thing to accept natural order as a way of life, but the minute one asks "why?" then again enters God and all His glory.

Our space ventures have been only the smallest of steps in the vast reaches of the universe, and have introduced more mysteries than they have solved. Speaking for myself, I can only say that the grandeur of the cosmos serves only to confirm my belief in the certainty of a Creator. Finite man cannot begin to comprehend an omnipresent, omniscient, omnipotent and infinite God. In the final analysis, any effort to reduce God to comprehensible proportions beggars His greatness. I find it best to accept God, through faith, as an intelligent will, perfect in goodness, revealing Himself through His creation—the world in which we live.

Science has taught us one most important lesson about God that we should never forget: we have learned that God does not interfere in the free order of life and nature which He created. If we do not accept this, we must abandon the entire concept of freedom.

The discoveries in astronomy, biology, physics, and even in psychology, have shown that we also have to enlarge the medieval image of God. If there is to be a mind behind the immense complexities of the multitude of phenomena which man, through the tools of science, can now observe, then it is that of a Being tremendous in His power and wisdom. But we should not be dismayed by the relative insignificance of our own planet in the vast universe as modern science now sees it. For it is perfectly conceivable that such a divine Being has a moral purpose which is being worked out on the stage of this insignificant planet.

When man, almost 2,000 years ago, was given the opportunity to know Jesus Christ, to know God who had decided to live for a while as a man amongst fellow men, on this little planet, our world was turned upside down through the widespread witness of those who heard and understood Him.

The same thing can happen again today. I am not in despair about the discordant conditions of our social environment. In spite of all the temporary setbacks that humanity has suffered through the centuries, and the terrible things that happened in our times, I strongly believe that God, in the same personal relationships He established through Jesus Christ, will see to it that man's path will continue upward, leading toward gradual improvement.

Jesus greatly expanded the basic moral laws of Judaism which formed the foundation of His teachings. His commandment to "love thy neighbor as thyself" established the unselfish attitude that enables human beings to live peaceably together. It is also the basis of our present foreign aid program. Even more revolutionary was His commandment to "love thy enemies." Although it is all too rarely followed, it has left an indelible and unforgettable imprint on the man-to-man relationships among people everywhere on our globe.

I am also confident that, as we learn more and more about nature through science, we shall not only arrive at universally accepted scientific findings, but also at a universally accepted set of ethical standards for human behavior.

This may sound like a highly optimistic statement in view of all the atrocities and acts of terror committed in our time. All right, so maybe I am an optimist. I am convinced that in spite of all temporary setbacks there is a slow but steady upward trend in mankind's universal standard of ethics.

Take the case of slavery. In antiquity, slavery was considered a perfectly normal thing. For the Greek and Roman civilizations it was an essential element in their way of life. The idea of developing a civilization without some people having to do the dirty work so others could write their poetry was considered absurd. Even during the early nineteenth century, the thought of operating a cotton plantation in the southern United States without slaves was considered impractical. Yet today the very concept of slavery is universally condemned and labeled as repulsive, even in totalitarian countries.

Science and technology undoubtedly made decisive contributions to the abolishment of slavery. They have provided everyone with a wide assortment of electrically- and gasoline-powered slaves, which once and for all did away with the need for any human slaves.

I am certain that this, too, was a part of the master plan of our Maker. In the world around us we can behold the obvious manifestations of His divine plan wherever we look. We can see the will of the species to live and propagate. We behold the gift of love. And we are humbled by the powerful forces at work on a galactic scale, and the purposeful orderliness of nature that endows a tiny and ungainly seed with the ability to develop into a beautiful flower. The better we understand the intricacies of the universe and all it harbors, the more reason we have found to marvel at God's creation.

Some of my scientific colleagues seem to have serious difficulties tying together certain Biblical passages with the reality of science, such as the story of creation given by Genesis, or the account of Joshua's poetic appeal for the Sun to stand still while the Israelites avenged themselves over their enemies. The interpretation of Biblical passages has been the subject of argument between wiser men than myself for centuries. My own views on the delicate topic are that it helps to bridge the gap between the Bible and modern scientific thought if we remember that the Bible deals with man as well as God, and most of the people of whom the Bible speaks suffered from the same human frailties that we experience today.

In my opinion, (and let me emphasize here that I fully respect and honor different views) insistence on an inflexible type of religion, holding to a literal interpretation of every word of the Bible as ultimate truth, will tragically delay reconciling some of the Biblical references to scientific interpretation. But I believe, with all my heart that religion, like science, is evolutionary, growing and changing in the light of further revelations by God. While the Bible is the best preserved account that we have of the revelations of God's nature and love, we should recognize that particularly the early books, such as Genesis, were not written by scientific observers and witnesses, but by scribes who recorded ancient shepherd songs and tales because of their allegorical beauty.

A scientist who discovers a new bit of knowledge does not tear down his model of reality, but merely changes it to agree with a new set of observations and experiences. By doing so, he admits he has no claim on ultimate truth. His laws are simply observations of reality, which he is always ready to update as his enlightenment grows.

And so, I think, it should be with the Christian religion. While preserving the ethical, moral, and spiritual meaning of the scriptures, the Christian churches should become a little more flexible with regard to various interpretations of the Bible as an historical account. We can still love and have faith in the words of the Bible, as they reveal so effectively so many time-honored truths.

The Christian churches cannot hope to reassume their rightful responsibility for ethical guidance with irrelevant debates concerning science versus religion.

Genetic engineering may serve as an example of a controversial and urgent issue about which many scientists would gratefully accept categorical guidance from the Church. The feasibility of manmade mutations by tampering with the famous double-helix-shaped genes which are imbedded in the chromosomes and determine the inheritable features of plants and animals, has already been repeatedly demonstrated. This capability opens the door to unfathomable opportunities for the breeding of high-yield strains of crops, straighter trees, or superior cattle, and we can expect to hear a lot of praise for all these potential benefits. But it seems that we are just one step away from being able to also tamper with the features of man.

Such a capability not only raises gruesome aspects of future dictators and madman breeding fearless soldiers or submissive citizens, it also touches on the fundamental issue of the dignity of man. It is here where, in my opinion, the Church should step in. It seems to me that agreement among Churchmen should be readily obtainable, as the Bible provides a clear and unambiguous answer. Man is a creation of God, who wanted him the way he is, with all his good and bad characteristics, and He wanted each of us as a responsible and a distinct individual. He has permitted him to probe His creation to his heart's content, but He never gave him the right to permanently deform or modify man whether as an individual or as a species.

But in imposing categorical guidance of this kind, the churches must be careful not to endanger the freedom of scientific pursuit. Although I know of no reference to Christ ever commenting on scientific work, I do know that He said, "Ye shall know the truth and the truth shall make you free." Thus I am certain that, were He among us today, Christ would encourage scientific research as modern man's most noble striving to comprehend and admire His Father's handiwork. I am certain that the leaders of our Christian churches know full well that they must come to grips with the world of Twentieth Century realities which constitute the main concern of contemporary man. The fresh wind which is blowing through most of the world's Christian institutions and meetings, like the one we are attending here, are sure indicators that the Churches are indeed responding to the new and unprecedented demands placed on them by the Space Age. Ecumenical Council meetings in Rome [and] interdenominational and Judeo-Christian conferences in the U.S. are strong evidence of a growing realization among Church leaders that the most important criteria for the future of our religious institutions, and the guiding role of Judeo-Christian ethics in these troubled times, is emphasis on the crucial issues of right and wrong selections in this complex scientific and technical world, that offers us more golden opportunities and more roads to disaster than any previous period in human history.

I am quite confident that the great majority of Christian Church leaders know in their hearts that this united front can best be presented by a common faith of all Christians in the basic teachings of Jesus Christ. But it means learning to live with the findings of Copernicus, of Galileo and Darwin. This front requires an emphasis on the essentials of spiritual life as identified by Jesus Christ rather than on trivia. It requires the acceptance of change and the discarding of antiquated or downright erroneous ideas, no matter how painful.

Personally, I noticed with deep regret the rejection by the Catholic Church of the ideas presented in *The Phenomenon of Man*, and some of his other books, by Teilhard de Chardin, the French Jesuit paleontologist who died about twenty years ago. I think Teilhard's books made an immense contribution to the reconciliation of scientific facts with religious beliefs.

His concept of a "convergent cosmogenesis" builds a solid bridge between Darwin's random evolution and survival of the fittest (both of which Teilhard accepts as elementary working mechanisms in God's workshop) and the obvious evidence of a unifying consolidation of the cosmos under the influence of the same supreme unity that laid down the basic, universal and all-embracing laws of nature. Teilhard's identification of the powers of love as the driving force behind this auto-convergence offers a beautiful explanation for the paradox that any upward development of the creation is in direct violation of the Second Law of Thermodynamics. This pitiless but all-embracing Law of Physics decrees that all random processes in nature lead from more order to more chaos; from states of lower probability (for instance, a higher degree of sophistication) to states of higher probability (i.e., a lower degree of sophistication); from a mountainous geography supporting a swift, energy-releasing stream to a sluggish river, incapable of giving off energy as it moves through the flat silt beds formed by water erosion of the same mountains that gave birth to the river. Since Darwin, who had never heard about mutations, we know that the vast majority of mutations on the genes of plant, animal and human life tend to deteriorate rather than improve a species. It's hard to see how with mutations, random evolution and survival of the fittest alone the living world on Earth could have progressed the way it did. But that doesn't mean we should throw Darwin's books out of the window.

The power of love, the most precious gift with which God endowed his creation, adds a new element to the dreadful probability statistics which, while still valid in the world of physics and in Darwin's evolutionary mechanism, all by itself can never account for the marvels of the world in which we live. Teilhard de Chardin tells us to view the universe as both lovable and loving. In so doing, he also interprets Christ's teachings on the overriding importance of love and charity as not just desirable virtues but as fundamental forces in the nature of the creation.

Siegmund Freud once remarked, "since Galileo, in the eyes of science, man has continuously lost, one after another, the privileges that had previously made him unique in the world. Astronomically, biological, psychologically ...." Now, paradoxically, Teilhard de Chardin finds that "this same man is in the process of reemerging from his return to the crucible, more than ever at the head of nature; since his very melting back into the general current of convergent cosmogenesis, he is acquiring in our eyes the possibility and power of forming in the hearts of space and time, a single point of universalization for the very stuff of the world."

To me, Teilhard de Chardin's ideas offer the most promising bridge between science and religion, and his formulations should be most satisfying to modern man. The rejection of his work was a particularly hard blow to Catholic scientists, but it hurt other believing scientists as well. I understand that the church rejected Teilhard because he left no place in his system for original sin. Now, while I am ready to confess that I am a sinner, and worse still, that I sometimes even enjoy being one, I must also confess that I have a hard time myself accepting the concept of original sin. None of us will probably live long enough to see it happen, but I am sure that Teilhard de Chardin will either be rehabilitated by the Catholic Church or go down in history as another Galileo.

In this reaching of the new millennium through faith in the words of Jesus Christ, science can be a valuable tool rather than an impediment. The universe as revealed through scientific inquiry is the living witness that God had indeed been at work. Understanding the nature of the creation provides a substantive basis for the faith by which we attempt to know the nature of the Creator.

After Frank Borman had returned from his unforgettable Christmas 1968 flight around the Moon with *Apollo 8*, he was told that a Soviet cosmonaut, recently returned from a space flight, had commented that he had seen neither God nor angels on his flight. Had Borman seen God, the reporter inquired? Frank Borman replied, "No, I did not see Him either, but I saw His evidence."

There can be no thought of finishing; for aiming at the stars, both literally and figuratively, is the work of generations; but no matter how much progress one makes, there is always the thrill of just beginning.

— von Braun

— Appendix —

Dr. Wernher von Braun was 50 years old on March 23, 1962. Sixty-four of his associates desired to show their affection by presenting him a birthday gift worthy of the event. It is customary in the scientific community to honor a respected master on his fiftieth birthday with a volume of papers written by those who have been privileged to work with him. What could be more appropriate for von Braun than a volume of technical and scientific papers on the subject which, for more than 30 years, had been so close to his heart?[106] Following are Dr. Ernst Stuhlinger's congratulatory remarks at von Braun's birthday party.

DR. VON BRAUN, MRS. VON BRAUN, HONORED GUESTS![107]

It has been customary in the scientific community for centuries that a respected and beloved master was honored on his birthday by a volume containing papers written by those who, at one time or another, were privileged to work for this master. Some achieve this state of perfection on their 75th birthday, others on their 100th, and still others on their 500th. In your case, Dr. von Braun, even your 50th birthday seems belated for this proof of respect and affection by your associates.

Last spring, almost a year ago, the plan for such a birthday volume was born. Many of your old and new friends desired to contribute, and the main problem was to keep the numbers of papers within manageable limits. I must say, though, that the original number of the contributors decreased somewhat, simply because rocket engineers have a habit of receiving, every once in a while, a TWX from up north saying, "mail us complete project development plan for Saturn family with alternate upper stages consisting of chemical, fission, fusion, ion, and photon engines by Friday next week." An assignment like this, of course, eliminates any extracurricular functions, and several of the prospective authors had to withdraw, very much to their regret, from contributing to this book.

Mentioning thermonuclear fusion rockets, by the way, reminds me of a little joke that Dr. Edward Teller told us last week in Berkeley. Talking about the Sherwood project, he asked whether anyone knew where the name came from. Nobody did, of course, so he explained it. The project is called Sherwood because it sher wood be wonderful if it ever worked!

This book comes to you today in a hard-bound edition. It was planned, written, edited and printed in house, and it came off the pilot manufacturing line in the true high-efficiency style which is characteristic of von Braun team projects. All of this happened, though, after permission and approval had been requested and obtained from almighty 801 19th Street. We were most gratified to receive a very fine prologue for the book from your newest boss, Mr. Webb, and an equally fine epilogue from your oldest boss, Dr. Dornberger.

The first edition will soon be followed by a second edition, bound in leather and cut in gold, to be published by McGraw-Hill. There will even be some royalties accruing from that book, and we have suggested that you may wish to donate these royalties in some appropriate way to the advancement of an educational institution. We know how highly you think of education, and how greatly you are interested in the development of educational opportunities. It is always nice if a man can put a little money behind the words that come from his heart.

Besides honoring you with this book on your birthday, we tried to pursue a second purpose with it. Everybody in the country knows you from the TV screen, and from your indefatigable speaking tours to bar associations, legislators, or newspaper men. Not everybody knows, however, that you are in the first place a rocket engineer, and that there is, as a characteristic byline of your team, sometimes even a modest touch of scientific endeavor at the Marshall Space Flight Center. Here we hope, the McGraw-Hill book may do a little good.

There are still many underprivileged persons around who can see you only extremely rarely. Among them are people who live overseas, some of your division directors, and Mrs. von Braun. For their benefit, we have included a photograph of you in the book. It was taken only a few weeks ago, and yet, you look as if it were taken twenty years ago in Peenemünde. I wish to say, though, that Dr. von Braun looked differently at that time. The ratio of payload to takeoff mass was smaller with rockets at that time, but considerably larger with humans.

The editing work on this book was done by three of your close associates: George Bucher, Fred Ordway, and Jerry McCall. George has a marvelous gift for coaxing others into abiding by their promises, and into exhibiting their best possible performances. He convinced the authors that they were working against a non-slipping deadline, and that the target date, in effect, was set and settled almost 51 years ago. Fred possesses an admirable feeling for everything that has to do with books and with writing. To me, it was almost a miracle to see how, under his hands, a gigantic pile of loose sheets, hand sketches, equation numbers, and paper clips transformed into a book, and how this book was ready a full week ahead of time. Jerry with his sharp-witted judgment and his outstanding sense of scientific quality, was the conscience of the editorial board. Besides this, he fulfilled a most valuable and necessary function in shielding you off against what was happening in your immediate vicinity, and what we wished should come as a surprise to you.

This beautifully bound McGraw-Hill book is still empty. This should not mean that you have discovered now a second place where the vacuum comes in cheap. It simply means that the printed version will come out soon. We would like to suggest that each of us should write his name in this book today, as a memento for Dr. von Braun of this birthday celebration. We would like to tell you, Dr. von Braun, that we did send an identical copy of the hand-bound book to your father. He should have it in his hands today.

In handing this volume over to you now, Dr. von Braun, we wish to express the hope that your next fifty years may bring you the fulfillment of your boldest dreams: a flight to Mars for you, and for all the rest of us your safe return to Earth.

— Notes —

1. Ernst Stuhlinger and Frederick I Ordway III, *Wernher von Braun: Crusader for Space*, Malabar, FL: 1996, p 95.
2. Dan O'Hair, *A Speaker's Guidebook,* Boston: 2001 p 263.
3. *X + 60 and Counting* is a book of letters from colleagues, both personal and professional, written to Wernher von Braun in celebration of his 60th birthday, March 23, 1972
4. Jules Verne, *From the Earth to the Moon,* New York: 1978 p 105
5. Crusader p 10
6. Ibid., p 11
7. Ibid., p 15
8. Ibid., p 17
9. Wernher von Braun and Frederick I Ordway III, *The Rockets' Red Glare,* New York: 1976, p 123.
10. Red Glare. p 128
11. Crusader p 33
12. James Harford, *Korolev,* New York: 1997, p 131.
13. Stuhlinger, p 61.
14. Ordway, p 16.
15. *X + 60 and Counting*
16. Stuhlinger, p 93.
17. *X+60 and Counting*
18. *X + 60 and Counting*
19. *X + 60 and Counting*. Bob Ward wrote in *Dr. Space*, "The first major breakthrough in spreading the space gospel to the American people came via the popular magazine *Collier's.*" There were eight articles published in the magazine—1952-1954, justifying that orbiting satellites and voyages to the Moon and back were achievable.
20. Stuhlinger, p 82.
21. "Future Development of the Rocket." El Paso, Texas. 1947. Box 101-3
22. Irene Powell-Willhite, "The British Interplanetary Society: Val Cleaver and Wernher von Braun," London: p 294.
23. *X + 60 and Counting*
24. Stuhlinger, p 109.
25. "The Importance of Satellite Vehicles in Interplanetary Flight" Read by Frederick Durant at the second Astronomical Congress in London, 1951. Box 149-5.
26. Powell-Willhite, p 294.
27. Stuhlinger, p 4.
28. Powell-Willhite p 294
29. Stuhlinger, p 92
30. "We Need a Coordinated Space Program," read by Frederick Durant at an IAF meeting in Zurich, 1953. Box 101-19.
31. The name A-4 was given to the Peenemünde rocket during development. The A-4 was a part of the evolving family of rockets. The name V-2 was introduced by Hitler's propaganda minister.
32. Paul Dickson, *Sputnik: The Shock of the Century*, New York: 2001, p 1.
33. John C. Lonnquist and David Winkler, *To Defend and Deter: The Legacy of the Cold War,* Rock Island, IL, 1996, p 1.
34. Michael Patterson, ed., *Major Problems in American Foreign Policy Volume II: Since 1914,* Lexington, Massachusetts: 1984, p 423.
35. Ibid., p 282.
36. Walter Issacson & Evan Thomas, *The Wise Men: Six Friends and the World They Made,* New York: 1986, p 380.
37. Dickson, p 1.
38. *X + 60 and Counting*
39. Robert Divine, *Sputnik Challenge* New York: 1993 p xiii.
40. Harford, p 121.
41. Divine, p xiv.
42. J.D. Hunley, Editor, *Birth of NASA: The Diary of T.Keith Glennan,* Washington, D.C.: 1993, p xix.
43. William Walter Space Age p 80
44. Divine, p xv.
45. Divine, p xvi.
46. von Braun, p 149.
47. "Space Superiority as a Means for Achieving World Peace," Box 101-12

48. X + 60 and Counting
49. "Voyage to the Moon," Box 101-13
50. No title, Box 104-6
51. "Youth Faces Tomorrow," Box 104-8
52. "Woman's Role in a Changing World," Box 104-7
53. "Patriotism in the Space Age," Box 104-9
54. No title Box 105-12
55. No title Box 105-2
56. No title Box 105-1
57. "The Struggle for the Future," Box 105-3
58. "Space, Its Problems, and You," Box 105-10
59. No title Box 149-7
60. Rocket Team p 260
61. No title Box 108-5
62. No title Box 108-4
63. No title Box 108-14
64. Sagan, p 333.
65. Ibid.
66. No title Box 111-15
67. Stuhlinger, p 271.
68. Immortality, Box 151-5
69. No title Box 115-9
70. Stuhlinger, p 146.
71. "Catch an Exploding Star," Box 113-5
72. No title Box 151-16
73. "The Making of a Missile," Box 113-1
74. "The Meaning of Space Exploration," Box 115-4
75. The early Saturn designations were C-1 etc. rather than S-1 etc.
76. No title Box 115-11
77. No title Box 115-16
78. Stuhlinger, p 170.
79. No title Box 115-15
80. "Shape Progress Report," Box 115-17
81. The Apollo illustrations, depictions of the proposed space vehicles, were used in von Braun's speech.
82. "Peaceful Uses of Space," Box 116-7
83. No title Box 116-8
84. No title Box 116-19
85. No title Box 117-5
86. No title Box 117-11
87. "The Challenge of the Century," Box 151-17
88. No title Box 118-8
89. Welcoming remarks, Box 129-21
90. Benson, p 476.
91. "Lunar Landing Celebration," Box 151-18
92. Stuhlinger, p11.
93. No title Box 149-9
94. "Religious Implications of Space Exploration," Box 143-16
95. No title Box 142-21
96. "Fall-Out Effects of Megascience" Box 143-2
97. "Face to Face Q & A," Box 109-28
98. "New Answers—New Questions," Box 144-12
99. "Benefits From Space and Space Technology," Box 144-11
100. Stuhlinger, p 311
101. No Title Box 150-19
102. Von Braun papers Box 144-16
103. Ward, p 217.
104. Stuhlinger, p 329.
105. "Responsible Scientific Investigation and Application," Box 134-15
106. From Peenemünde to Outer Space MSFC NASA Ernst Stuhlinger, Ordway, McCall, Bucher p x
107. Box ES103-16 Ernst Stuhlinger collection
108. Francis B. Sayre, Jr. Memorial Service Wernher von Braun 1912-1977 Washington Cathedral June 22, 1977
109. Box ES101-13 Ernst Stuhlinger collection

— Bibliography —

X+60 and Counting, is a book of letters, personal and professional, written to Wernher von Braun in celebration of his 60th birthday, March 23, 1972.

Benson, Charles D., Faherty, William Barnaby, *Moonport: A History of Apollo Launch Facilities and Operations*. Washington, D.C.: National Aeronautics and Space Administration, 1978.

Dickson, Paul, *Sputnik The Shock of the Century*. New York: Walker & Company, 2001.

Divine, Robert A., *The Sputnik Challenge*. New York: Oxford University Press, 1993.

Harford, James, *Korolev*. New York: John Wiley & Sons, Inc, 1997.

Hunley, J.D., Editor, *The Birth of NASA: The Diary of T. Keith Glennan*. Washington, D.C.: National Aeronautics and Space Administration, 1993.

Isaacson, Walter and Thomas, Ivan, *The Wise Men: Six Friends and the World They Made*, New York: Simon and Schuster, 1986.

Lonnquest, John C. and Winkler, David F, *To Defend and Deter:The Legacy of the United States Cold War Missile Program*. USACERL Special Report 97/01 November 1996 Department of Defense Legacy Resource Management Program, Cold War Project Defense Publishing Service, Rock Island, IL.

O'Hair, Dan, *A Speakers's Guidebook*. Boston: St. Martin's, 2001.

Ordway, Frederick I, III and Sharpe, Mitchell R., *The Rocket Team*. Ontario, Canada: Apogee Books, 2003.

Paterson, Thomas, Editor, *Problems in American Foreign Policy*. Lexington, Massachusetts: D.C. Heath and Company, 1984.

Sagan, Carl, *Cosmos*. New York: Random House, 1980.

Stuhlinger, Ernst, Ordway III, Frederick I, *Wernher von Braun: Crusader for Space*. Malabar, FL: Krieger Publishing Company, 1996

Verne, Jules, *From the Earth to the Moon*. New York: Gramercy Books, 1978.

von Braun, Wernher and Ordway, Frederick I, III, *History of Rocketry & Space Travel*. New York: Thomas Y. Crowell, 1966.

von Braun, Wernher, and Ordway, Frederick I, III, *The Rockets' Red Glare*. New York: Anchor Press/Doubleday, 1976.

Ward, Bob, *Dr. Space: The Life of Wernher von Braun*. Annapolis, Maryland: Naval Institute Press, 2005.

Journal: Powell-Willhite, Irene, "The British Interplanetary Society: Val Cleaver and Wernher von Braun," JBIS (2001): 291-299.